Electromagnetic Non-Destructive Testing and Evaluation

Electromagnetic Non-Destructive Testing and Evaluation

Guest Editor
Yong Li

MDPI

Basel • Beijing • Wuhan • Barcelona • Belgrade • Novi Sad • Cluj • Manchester

Guest Editor
Yong Li
School of Aerospace
Engineering
Xi'an Jiaotong University
Xi'an
China

Editorial Office
MDPI AG
Grosspeteranlage 5
4052 Basel, Switzerland

This is a reprint of the Special Issue, published open access by the journal *Sensors* (ISSN 1424-8220), freely accessible at: https://www.mdpi.com/journal/sensors/special_issues/U8C667CQ0X.

For citation purposes, cite each article independently as indicated on the article page online and as indicated below:

Lastname, A.A.; Lastname, B.B. Article Title. *Journal Name* **Year**, *Volume Number*, Page Range.

ISBN 978-3-7258-5563-6 (Hbk)
ISBN 978-3-7258-5564-3 (PDF)
https://doi.org/10.3390/books978-3-7258-5564-3

Contents

About the Editor

Yong Li

Yong Li has been with the State Key Laboratory for Strength and Vibration of Mechanical Structures, Shaanxi Engineering Research Centre of NDT and Structural Integrity Evaluation, School of Aerospace Engineering, Xi'an Jiaotong University, as a full Professor in electromagnetic non-destructive testing and evaluation. He received his B.Sc. degree and M.Sc. degree in electrical engineering from Xi'an Jiaotong University, Xi'an, China, in 2001 and 2004, respectively, and his Ph.D. degree from Newcastle University, Newcastle upon Tyne, U.K., in 2009. His research interests involve electromagnetic non-destructive testing and evaluation, forward modelling and inversion for defect reconstruction, microwave reflectometry for imaging, the characterisation and evaluation of structural defects, 3D magnetic field sensing and gradient field measurement, image/signal processing and feature extraction techniques, etc. He has coordinated several research projects from the National Natural Science Foundation of China.

sensors

MDPI

Article

Resonant Eddy Current Sensor Design for Corrosion Detection of Reinforcing Steel

Upeksha Chathurani Thibbotuwa [1,*], Ainhoa Cortés [1,2,*], Aurora María Casado [3] and Andoni Irizar [1,2]

1 CEIT-Basque Research and Technology Alliance (BRTA), Manuel Lardizabal 15, 20018 Donostia-San Sebastián, Spain; airizar@ceit.es
2 Department of Electronics & Communications, Tecnun, Universidad de Navarra, Manuel Lardizabal 13, 20018 Donostia-San Sebastián, Spain
3 ACCIONA Construction Innovation Technology Division, C/Valportillo II no 8, 28108 Alcobendas, Spain; auroramaria.casado.barrasa@acciona.com
* Correspondence: uthibbotuwa@ceit.es (U.C.T.); acortes@ceit.es (A.C.); Tel.: +34-943212800 (U.C.T. & A.C.)

Abstract: This paper introduces an LC resonator-based single-frequency eddy current (EC) sensor designed for corrosion detection in reinforcing bars (rebars) embedded within concrete structures. The work addresses the challenges of the limited detection ranges and reduced sensitivity over longer distances, prevalent in current EC sensor applications. The sensor development process involved a systematic experimental approach to carefully selecting each parameter in the LC resonator. The sensor design aimed to assess the condition of the rebar from a distance of up to 5–6 cm outside the concrete and provide insights into different corrosion levels. By examining the characteristics of the inductors, the parallel resistance R_p of the eddy current coil was identified as a key parameter reflecting the corrosion conditions in the rebar. The relationship between the R_p fluctuations and temperature variations was investigated, with the data indicating that an approximately 155 Ω variation can be expected per 1 °C change within the temperature range of 20–25 °C, allowing for temperature compensation if necessary. Subsequently, the sensor's performance was evaluated by placing a rebar within a concrete block, where controlled mechanical degradation cycles were applied to simulate uniform corrosion in the rebar. The experimental results show that our EC sensor can detect material loss around the rebar with accuracy of approximately 0.17 mm.

Keywords: corrosion detection; eddy current; LC resonator; reinforcing bar; single frequency

1. Introduction

Reinforced concrete structures are susceptible to corrosion, a natural process that initiates when the steel rebar within the concrete begins to corrode. Corrosion in reinforced concrete occurs when the protective layer around the steel reinforcement is compromised, making the steel vulnerable to deterioration. The consequences of corrosion can be severe, leading to a reduced load-bearing capacity, structural instability, and potentially catastrophic failures if left untreated. Given the hidden nature of corrosion within concrete structures and the potential for rapid deterioration once visible, early detection is crucial for timely maintenance and in preventing costly damage. However, corrosion detection in reinforced concrete structures at an early stage presents challenges due to the concealed nature of the reinforcement bars and the complex interaction between the corrosion process and the concrete matrix. These challenges become more pronounced in offshore environments due to the combined environmental effects of moisture, saltwater, and oxygen, which accelerate the corrosion process. Moreover, offshore settings introduce further complications in inspection and maintenance activities due to restricted accessibility, adverse operational conditions, technological constraints, and costs.

Focusing on the reinforcing bars (rebars) within concrete elements, corrosion in these rebars is essentially an electrochemical process that occurs through a series of chemical reactions involving the steel (i.e., reinforcing steel) and the moisture and oxygen present in the concrete environment (i.e., concrete pore solution). Once the corrosion of the reinforcement steel in a reinforced concrete structure begins, it tends to advance at a relatively constant rate [1], and, with time, the corrosion products (iron oxides and hydroxides) are deposited in the limited space of the concrete surrounding the steel. This will cause expansive stresses, leading to cracks and spalling the concrete cover.

Reinforced concrete (RC) structures can experience both uniform and non-uniform corrosion, depending on various factors and the specific conditions to which they are exposed. Uniform corrosion typically occurs under relatively favorable environmental conditions, where the concrete structure is subjected to a consistent and mildly corrosive environment [2]. Uniform corrosion in atmospherically exposed RC structures is influenced primarily by the availability of water and the pore structure at the steel–concrete interface. On the other hand, non-uniform corrosion is characterized by the uneven or localized corrosion (such as pitting) of the steel reinforcement. This type of corrosion is often more concerning and damaging than uniform corrosion. Non-uniform corrosion can result from various factors, including localized exposure, the presence of chlorides, the quality of the concrete mix (concrete properties), dissimilar metals, and local environmental factors such as the micro-climatic conditions—areas with standing water or exposure to pollutants can contribute to localized corrosion [2,3]. Additionally, the temperature and humidity can cause both uniform and non-uniform corrosion in reinforced concrete (RC) structures, primarily through their influence on the moisture content within the concrete and the mean temperature of the concrete structure [4].

Non-destructive testing (NDT) techniques are widely employed to investigate the mechanisms and kinetics of corrosion in steel embedded within concrete structures. These NDT methods are invaluable in evaluating the condition of reinforcing steel bars (rebars) without exerting any destructive impact on the test object. For long-term corrosion monitoring systems, it is recommended to employ technological approaches that are both NDT and non-invasive, where the non-invasiveness means that the testing method does not alter the material condition due to the technical approach that we use. Some of these techniques focus on determining the concrete properties, while others focus on the rebar properties for corrosion detection.

- Concrete Property Evaluation: This primarily focuses on assessing the properties of the concrete itself and helps to determine how susceptible the concrete is to corrosion. It may include measures of the concrete's porosity, moisture content, and permeability and the presence of cracks or defects. Understanding these concrete properties is crucial because they influence the ingress of aggressive substances, such as chloride ions or carbon dioxide, which can trigger steel corrosion within the structure. Some examples are surface potential (SP) measurements, concrete resistivity measurement, infrared thermography, etc.
- Rebar Property Analysis: On the other hand, some techniques are more focused on evaluating the properties of the reinforcing steel bars (rebars) within the concrete. These methods aim to assess the condition of the rebar, such as its corrosion rate, corrosion products, and thickness loss and the degree of steel passivation. Understanding the rebar properties is essential because it allows us to directly assess the corrosion condition of the rebars. Some examples are half-cell potential (HCP) measurement, open circuit potential (OCP) measurement, linear polarization resistance (LPR) measurement, ultrasonic testing (UT), electromagnetic testing, etc.

In this context, the corrosion detection sensors employed in NDT techniques play a key role in the timely assessment of corrosion within reinforced concrete by providing the early detection of damage [5]. As these sensors measure parameters indicative of corrosion, the resolution and sensitivity that they offer are important in accurately reflecting the current state of corrosion. However, these sensors have inherent limitations based on their sensing principles, which may become particularly challenging based on the application requirements. For example, when rebars embedded in concrete are inspected from the outer surface, the use of ultrasound sensors presents more challenges compared to that of electromagnetic sensors. This is mainly because ultrasound sensors require extra preparation to establish good-quality contact with the testing concrete surface; furthermore, high-frequency ultrasound waves are significantly attenuated in thicker concrete structures, leading to the potential loss of critical information. Additionally, the internal structure of concrete, with its varying density and the presence of aggregates, can also cause the scattering and reflection of the ultrasound waves, further complicating data interpretation. In contrast, electromagnetic sensors are primarily responsive to conductive materials. Provided that the magnetic field that they generate is strong enough, they can be more effective in assessing only the rebar within the concrete, as they are less influenced by the surrounding non-conductive matrix.

The assessment of corrosion in reinforced concrete structures has been the subject of extensive research, leading to the development of various sensor-based techniques. Numerous reviews in this field offer insights into the different assessment methods [3,6,7]. Since the corrosion of reinforcement steel in concrete is an electrochemical process, electrochemical methods have traditionally been the most prevalent techniques used to assess and determine the corrosion rate in concrete structures.

However, there is a growing trend towards evaluating corrosion in reinforced concrete by examining its physical properties. These physical property assessment techniques offer several advantages, which include the provision of immediate results during testing (measurement speed), allowing for real-time assessments, offering a direct and multi-parameter approach to detect various corrosion-related issues, and including the thickness loss, reduced adhesion at the steel/concrete boundary, and cracks due to corrosion product accumulation. While electrochemical methods remain important, the industry increasingly recognizes the benefits of physical property assessment techniques for more effective and timely corrosion monitoring practices.

The approach proposed in this paper focuses on assessing the physical parameters of rebars for corrosion monitoring. Specifically, this solution aims to monitor concrete offshore structures (e.g., breakwater) to protect photovoltaic (PV) offshore platforms. Within this framework, our efforts are dedicated to developing a robust and cost-effective corrosion monitoring sensor node to be deployed on offshore breakwater.

2. Sensor Development: Corrosion Detection in Rebars by Physical Property Assessment

Critical factors for the selection of an appropriate sensing principle include ensuring the sensitivity to detect developing corrosion at earlier stages and the selectivity to distinguish corrosion from other interfering signals, such as environmental noise, physical disturbances, temperature variations, and humidity changes. Moreover, given the objectives of the application, it is important to consider the required measurement depth and ensure compatibility with the intended sensor integration and data collection infrastructure. Moreover, in offshore environments, the sensor must be durable, withstanding harsh marine environmental conditions, and provide reliable data. The practical usability of such sensors is underpinned by straightforward calibration procedures, user-friendly interfaces, and sophisticated algorithms for accurate data analysis. Lastly, regulatory compliance, cost-effectiveness, and sustainability considerations are also important in successfully implementing these sensors in the construction and inspection industry.

In the context of evaluating steel bars within reinforced concrete structures, a non-destructive technique such as eddy current testing (ECT), which is based on electromagnetic induction, enables the assessment of the reinforcement's condition without requiring invasive procedures. This method involves generating localized circular currents, known as eddy currents, within the steel bars through the application of electromagnetic fields. The subsequent analysis of variations in these currents helps in detecting discontinuities or changes in the material properties of the steel bars, such as potential corrosion or cracks. When applied to offshore concrete structures, electromagnetic NDT methods offer several advantages. These include the elimination of direct physical contact with the rebars or concrete structure, which is particularly beneficial in environments where access may be constrained. Additionally, these methods generally require minimal surface preparation, which can greatly reduce the inspection preparation times by eliminating the need for coating removal or extensive cleaning beforehand. This facilitates a rapid inspection process, enabling quick measurements and the potential for the real-time monitoring of structural health. Furthermore, electromagnetic NDT methods can provide quantitative data on the condition of the rebars, such as their level of corrosion or the loss of the cross-sectional area, which is important for the effective maintenance and extension of the service life of offshore structures.

In our application, we anticipate the development of a compact, low-power, and cost-effective sensor node that can be used to assess the condition of rebar corrosion from the exterior of a concrete structure. Given these specifications, eddy current NDT emerges as a particularly fitting technique for the detection of corrosion in reinforcements. Its non-destructive nature and its ability to detect surface and sub-surface flaws in conductive materials embedded in concrete, without requiring direct contact with the surface of the concrete, is very suitable for the target application. Moreover, it enables the low-cost sensing of conductive targets in the presence of dust and moisture, making this technology reliable in harsh environments.

This area of study aims to explore the various aspects of the eddy current technique concerning corrosion-induced degradation in concrete-based infrastructure, including the following.

- Eddy current principles: studies have explored the fundamental principles of eddy current testing, which involves inducing electromagnetic currents in the steel bars and analyzing the changes in the eddy current properties caused by corrosion.
- Sensor development: studies have focused on developing specialized sensors and probes capable of generating and measuring the eddy currents in steel bars.
- Corrosion detection: research has aimed at refining the sensitivity of eddy current techniques to detect and quantify corrosion in steel reinforcements.
- Data analysis: studies have developed algorithms and data analysis methods to interpret the signals generated by eddy current testing.
- Field applications: field trials and case studies have been used to evaluate the effectiveness of eddy current-based corrosion monitoring in real-world scenarios.
- Long-term corrosion monitoring: given the importance of long-term corrosion monitoring in infrastructure maintenance, studies have explored strategies for the continuous and remote monitoring of reinforced steel bars using eddy current technology.

State of Art: Eddy-Current-Based Corrosion Detection

Eddy current (EC) probe design is essential in obtaining good results in eddy current testing. There are different types of eddy current probes available, such as single coil probes, eddy current array probes, rotating probes, bobbin probes, flexible probes, etc. A review of the eddy current testing probes for carbon steel pipeline assessment inspection is presented in [8]. Depending on the probe design, eddy current testing can be performed using either a single frequency or multiple frequencies. Single-frequency eddy current probes are designed to operate at a fixed, predetermined frequency during the inspection process. These probes are particularly quick to identify the presence or absence of defects or for thickness estimation [9]. Multi-frequency eddy current sensors use several frequencies to assess the test object. A multi-frequency approach in the estimation of the thickness for a given material is discussed in [10]. This work analyzes the change in the peak frequency of the imaginary part of the inductance using a non-magnetic metallic plate.

The effectiveness of the eddy current sensor is closely related to the precise design of the inductor coil, as it plays a crucial role in determining their performance and functionality. The coil's geometry, encompassing factors like its shape and size, directly impacts the magnetic field and its interaction with the target material [11,12].

An LC tank-based eddy current corrosion detection system is a cutting-edge technology that employs an LC (inductance–capacitance) resonator as its measurement probe. In eddy current NDT-based applications, an LC resonator has been adapted to serve as a single-frequency probe in eddy current testing. In general, an LC resonator is designed to resonate at a pre-defined frequency, and this fixed resonance frequency is utilized as the probing frequency during the inspection.

In the literature, there are several research works found based on ECT with LC resonators, considering different scopes. The authors of [13] propose a method to characterize the conductivity measurement of non-ferromagnetic metallic materials using an LC probe. The measurement parameters—the resonant frequency and series resonant resistance of the coil—are modeled as a function of the conductivity and the lift-off effect, where an experiment is carried out to test samples with different electrical conductivity values. In certain configurations, eddy current sensors are designed with separate transmitter and receiver coils, as discussed in the work by Andringa (2005) [14]. This work presents a prototype of a sensor with an LC resonator developed for reinforced concrete that consists of a reader and LC resonator circuit, which are magnetically coupled. A bare steel wire is connected to the LC resonator circuit to act as a transducer by exposing it to the same concrete environment as the reinforcing steel. The phase response of the capacitor of the resonator circuit is measured to read the corrosion activity of this bare steel wire. In this study, instead of directly measuring the corrosion condition of the rebars, they assess it indirectly by subjecting a bare steel wire to the concrete environment. A self-frequency conversion measurement system is presented in [15] based on an LC resonator for fiber-reinforced plastics in its conventional frequency range of operation of 1 kHz to 1 MHz. The measurement system is designed so that the resonant frequency is not fixed, allowing it to change yet keeping it always at a resonant level. A controlled oscillator with a feedback loop is used to detect the change in the resonant state. In [16], a three-point capacitive resonance circuit is presented for defect inspection in carbon-fiber-reinforced polymer testing based on the change in the resonance frequency of the output signal.

Existing EC sensors often encounter challenges related to their restricted detection ranges and low sensitivity, particularly when operating over longer distances. To our knowledge, achieving reliable detection at distances around 5 cm is uncommon in current applications, as eddy currents are typically utilized for shorter-distance inspections. Addressing this challenge, our research focuses on developing an EC sensor designed to detect different corrosion levels in reinforcing bars embedded in concrete and operates effectively over longer distances. For this task, we use an experimental methodology focused on sensitivity and robustness, quantified corrosion measures, specifically target applications where precise monitoring at greater distances is essential. Our research represents a significant step forward in EC sensor technology, offering a specialized solution for reinforced concrete structures that meets the specific demands of corrosion detection from comparatively greater distances.

3. Eddy Current Testing Principle

Eddy current testing (ECT) for corrosion detection in rebars operates on the principle of electromagnetic induction. When an alternating current is passed through a coil or probe, it generates an alternating magnetic field around it. If an electrically conductive material is in the vicinity of this electromagnetic field, surface currents known as eddy currents are induced within the material, according to Faraday's law of electromagnetic induction. The magnitude and distribution of the eddy currents are influenced by the distance, size, and composition of the conductor. These induced eddy currents, in turn, create an electromagnetic field opposing the primary magnetic field produced by the coil. The interaction between these fields induces a back electromotive force (EMF) in the coil, causing a measurable change in the coil's impedance.

When corrosion or defects are present within the rebar, they lead to variations in its material properties, particularly its magnetic permeability and electrical conductivity.

These variations lead to inconsistencies in the eddy current patterns during the ECT process. Corrosion typically results in lower electrical conductivity because it changes the metal into non-conductive corrosion products. This change is reflected as an alteration in the impedance of the eddy currents. In addition, the magnetic permeability of the rebar is influenced by the presence of corrosion. The development of corrosion products or the introduction of cracks can disrupt the uniform magnetic properties of the steel. As a result, the interaction of the probe's magnetic field with the rebar is changed, which in turn affects the secondary magnetic field produced by the eddy currents. By measuring and analyzing these disturbances, ECT can assess the condition of the rebar, detecting variations such as corrosion-induced pitting or thinning.

One major advantage of this technique is its ability to function without requiring direct contact with the test object, making it appropriate for online and in-service applications. However, for a long-term, reliable EC sensor, further studies are necessary to address challenges such as calibration complexities, the accurate quantification of corrosion, robust measurements, and the mitigation of environmental factors that affect the measurement accuracy.

Single-Frequency Eddy Current Sensor with an LC Resonator

LC resonators can function as eddy current probes operating at a single frequency by configuring them within a resonant LC circuit arrangement. This arrangement consists of a coil (inductor, L) and a capacitor (C) connected in parallel, forming a resonant LC circuit.

The eddy current coil can be modeled as an inductor L in series with a resistor R_s (see Figure 1), where the resistor R_s signifies the parasitic resistance of the coil. When an alternating current (AC) passes through the inductor, it generates an alternating magnetic field. In the presence of a conductive object near this magnetic field, the eddy current effect induces a change in L_s and R_s, which vary as a function of the distance (d) between the inductor and the object, as well as the characteristics of the conductor.

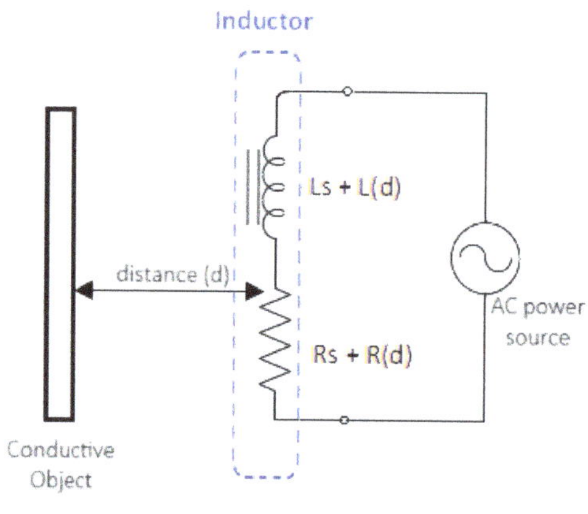

Figure 1. Inductor modeling with eddy current principle.

However, relying solely on an inductor to produce an alternating magnetic field demands a significant amount of power. As a solution, the power consumption is lowered by adding a parallel capacitor, forming an LC resonator. By carefully selecting the values of the coil and capacitor, the circuit is tuned to resonate at a specific frequency.

The LC circuit resonates at a frequency (f_0) given by Equation (1):

$$f_0 = \frac{1}{2\pi\sqrt{LC}} \tag{1}$$

where (f_0) is the resonance frequency, (L) is the inductance, (C) is the capacitance, and (π) is the mathematical constant Pi (approximately 3.14159).

When modeling an LC resonator, one common topology is the parallel R-L-C model, in which the resistor (R), inductor (L), and capacitor (C) components are all connected in parallel. For the purpose of simplifying the calculations related to the coil's amplitude, this parallel electrical model is preferred. In this model, the inductive component (L) represents the probe's coil, which is part of the eddy current probe assembly. The capacitive component (C) is typically a discrete capacitor that is selected or tuned to set the desired resonant frequency of the circuit. The resistive element (R) accounts for the parasitic losses, including the inherent resistance of the coil's windings and other inefficiencies that contribute to energy dissipation.

In an RCL parallel resonator, the detection of changes in a test object is facilitated through two primary methodologies: by tracking shifts in the resonant frequency and by examining fluctuations in parasitic resistance. The resonant frequency of the circuit is sensitive to variations in inductance and capacitance. Introducing a test object into the resonator's influence can cause a change in the coil's inductance, leading to a detectable alteration in the resonant frequency. This change can be quantitatively analyzed to reveal the object's position or other related characteristics.

On the other hand, it also offers a means to monitor the parasitic resistance of the circuit, which includes all non-ideal resistive losses, including those caused by the test object through mechanisms like eddy currents. Since the parasitic resistance influences the energy dissipation in the circuit, it affects the decay rate of the oscillations after excitation. By employing a feedback loop that continuously monitors the amplitude of the oscillations in the LC circuit, changes in the decay rate can be observed. These changes are indicative of variations in the parasitic resistance. The feedback loop adjusts either the excitation signal or the measurement parameters in real time based on the measured decay rate. An increase in the decay rate suggests an increase in parasitic resistance, and, conversely, a decrease in the decay rate suggests a decrease in parasitic resistance. Therefore, by analyzing the decay rate of the oscillations, the sensor can measure the parallel resistance of the LC circuit, which also includes the contributions from the parasitic losses from eddy currents. Since the eddy current induced in a conductor is influenced by factors such as its size, composition, and distance from the coil, both the parasitic resistance and the inductance of the coil fluctuate accordingly.

4. Experimental Methodology: EC Sensor Design

A systematic approach was followed in designing an EC sensor to meet the specific specifications and performance criteria required for our application, which primarily focused on detecting corrosion in the reinforcement bars of an offshore breakwater structure.

Accordingly, this paper presents a resonant EC approach utilizing a single frequency. The detection probe, referred to as the LC probe in this paper, is an LC resonator consisting of an eddy current coil and a capacitor connected in parallel. The sensor's development has been focused on two main objectives.

1. The sensor must evaluate the rebar condition from outside the concrete. Thus, the primary magnetic field generated should be strong enough to penetrate up to 5–6 cm (assuming that the concrete column thickness falls within this range).
2. The sensor should give meaningful information about the different levels of corrosion in the rebar.

Focusing on these main objectives, the approach to designing and implementing the sensing component (LC probe) is outlined through a series of pivotal steps as given below. The steps involved selecting suitable characteristics for the eddy current coil, followed by the selection of an appropriate capacitor to create the LC probe and subsequently testing the LC probe.

Step 1: Study the characteristics of the inductor/eddy current coil for the LC probe under predefined application requirements, considering the following.

- Self-resonance frequency: designing a probe to operate near its self-resonance frequency (SRF) can enhance its responsiveness and can be advantageous for certain applications, potentially leading to improved performance. However, beyond the SRF, the probe's inductive reactance diminishes, and it begins to exhibit capacitive characteristics, which affects its performance. Knowledge of the SRF is, therefore, important in designing LC probes that are effective and reliable across different operational scenarios.
- Physical size of inductor: the physical dimensions of an inductor contribute to its overall inductance, which in turn determines the intensity of the magnetic field that it can generate. Additionally, it is a consideration with regard to managing the weight and physical profile of the sensor system, particularly for applications where the sensor needs to be compact and easily deployable.
- Inductance value: the inductance (L) of an inductor is affected by its physical dimensions, including the number of turns in the coil, the cross-sectional area of the coil, and the core material. A larger inductor can have higher inductance, which can be advantageous in creating a stronger magnetic field or when working at lower frequencies, both of which can be beneficial for long-range sensing applications.
- Inductor shape: the shape of the inductor in an LC sensor determines the geometry of the electromagnetic field that it generates. A circular spiral shape is ideal for proximity applications where the target moves orthogonally to the sensor plane because it generates a more symmetrical magnetic field compared to other shapes [17].

Step 2: Choose the most suitable parameters for the interpretation of corrosion conditions. This involves testing inductors to study the variations in parameters such as the parallel resistance (R_p), series resistance (R_s), inductance (L_s), and impedance (Z) during a frequency sweep. The selection is based on each parameter's sensitivity to detect a conductive object (rebar) from a distance of 5 cm.

Step 3: Determine the operating frequency for the LC probe oscillation. This was determined by analyzing the frequency range that maximized the variation in the selected inductor's parameters (from Step 1) to optimally reflect the corrosion condition under predefined test conditions.

Step 4: Once the frequency of operation and inductance are chosen, select a suitable capacitor value for the LC probe. This is done by substituting the inductor's inductance value at the selected operating frequency range from Step 3 and calculating (using Equation (1)) the capacitor value so that the LC resonator's resonance falls within the same frequency range.

Step 5: Integrate the LC probe with an inductance-to-digital converter to test the digital output of the final sensor design.

Theoretically, for an EC sensor, a lower resonance frequency in the LC resonator circuit is preferable when deeper inspection depths are required. To facilitate this, a higher inductance value for the inductor is desirable, as per the fundamental resonance equation of an LC circuit (Equation (1)).

To assess various possibilities, we tested a selection of inductors that can be categorized into two types based on their physical properties and inductance: a circular inductor with inductance in the mH range and a few planar inductors that were smaller in size and with inductance values in the µH range. We will summarize the overall results obtained for each category using the results from two inductors as follows.

Inductor 1: an inductor with a 10 cm internal diameter and inductance of approximately 16 mH (see Figure 2a).

Inductor 2: an inductor developed for wireless power applications (Part No: 760308103145) by Würth Electronics with inductance of 33.9 µH, with a series configuration of three coils (see Figure 2b) [18].

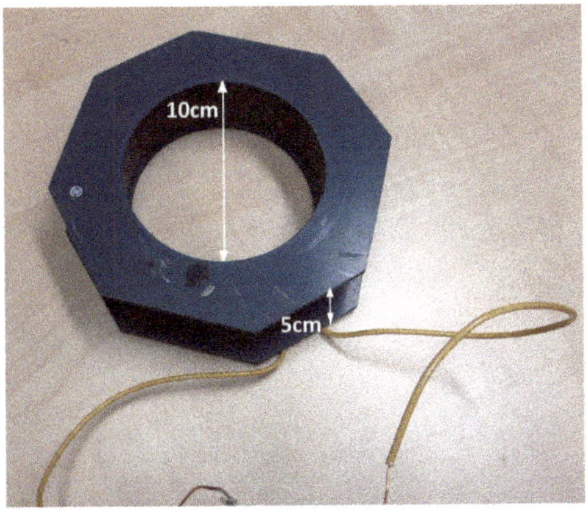

(**a**) Inductor 1: inductance of 16 mH

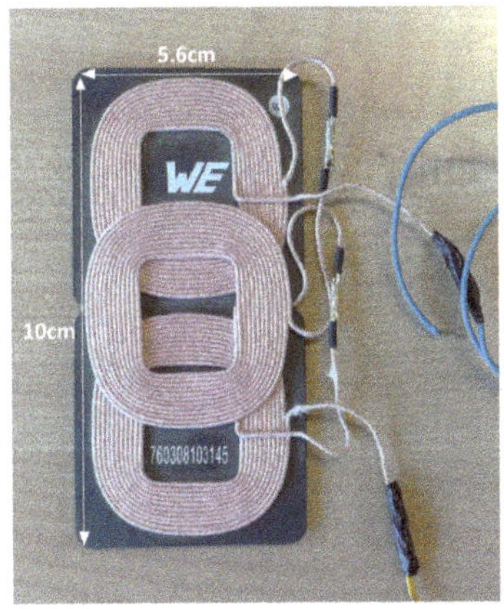

(**b**) Inductor 2: inductance of 33.9 μH

Figure 2. Photographs of Inductor 1 (**a**) and Inductor 2 (**b**).

A test setup was designed to allow adjustments to the height between the inductor and the non-corroded rebar for testing at various elevations, while simultaneously maintaining a constant distance between the sensor and the non-corroded rebar during measurements. This test setup was employed to evaluate the performance of the inductors under two conditions.

Condition 1: WO condition—with object condition, where a non-corroded rebar was positioned at a 5 cm distance within the vicinity of the inductor, as shown in Figure 3a;

Condition 2: NO condition—no object condition, where the rebar was removed from the inductor's vicinity, as shown in Figure 3b.

Based on our application, we assume that the WO condition with a non-corroded rebar represents 0% corrosion, and the NO condition in the absence of a rebar represents 100% corrosion.

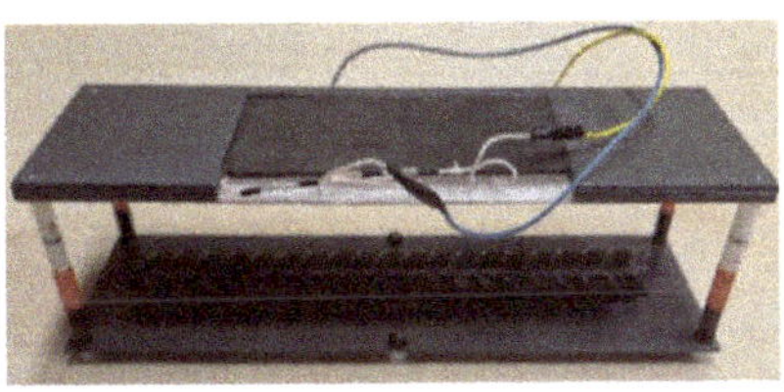

(**a**)

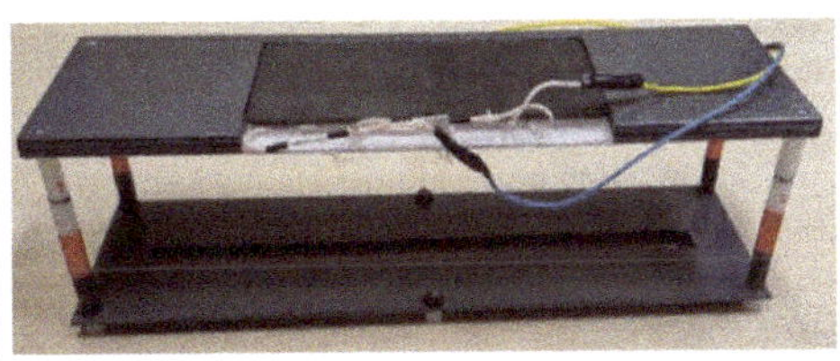

(**b**)

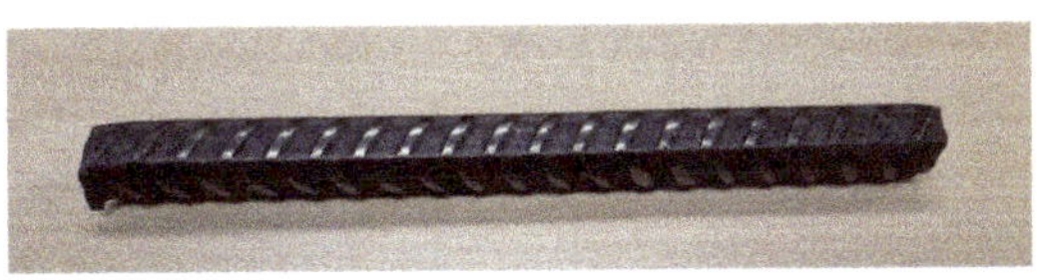

(**c**)

Figure 3. Testing condition demonstration (WO and NO) using Inductor 2 and the rebar sample S000 related to the LC probe design experiment. (**a**) Inductor 2: WO condition testing; (**b**) Inductor 2: NO condition testing; (**c**) Non-corroded rebar (S000) with 13 mm average diameter.

Following the design and implementation steps outlined in the beginning of this section, an experimental framework was established featuring a series of testing phases using two key pieces of test equipment: an LCR meter (HIOKI IM3536—Figure 4a) and an LDC1000EVM evaluation module from Texas Instruments (Dallas, TX, USA) (Figure 4b). The LCR meter was used to evaluate the inductor properties aligned with our application requirements. The main objectives were to determine the most appropriate parameters for the interpretation corrosion conditions, study the characteristics of the inductor design, and select a suitable operating frequency range for the LC probe oscillation.

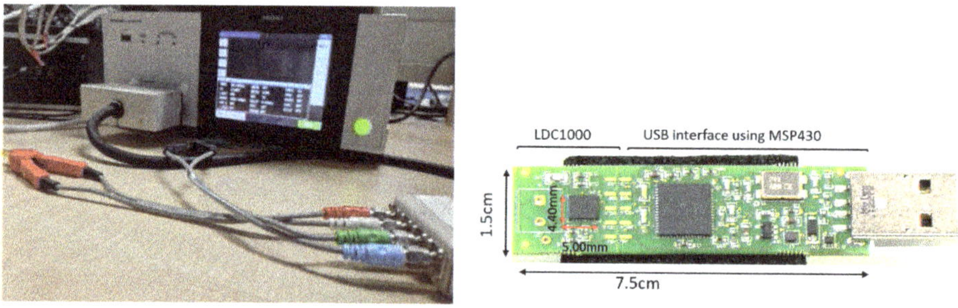

(**a**) (**b**)

Figure 4. Testing equipment used during the experiment. (**a**) HIOKI IM3536 (laboratory-level testing); (**b**) LDC1000EVM device.

Subsequently, the digital output produced by the LC probe was assessed by integrating the LC probe with the LDC1000EVM inductance-to-digital converter [19]. This test aimed to evaluate the digital output produced by the LC probe when interfaced with an LDC100x series IC, which will be incorporated into our final sensor node design.

In the following Sections 4.1 and 4.2, the methods, results, and key conclusions from the measurements conducted with the LCR meter and the LDC1000EVM are discussed, respectively.

4.1. Testing Inductors' Properties: Using LCR Meter

The inductors were positioned on the test setup with their terminals connected to the measurement probes of the LCR meter for a frequency sweep analysis under both WO and NO conditions. During the WO condition testing, the non-corroded rebar S000 was placed in the setup at a fixed distance of 5 cm from the inductor, as illustrated in Figure 3a.

In the data analysis, the percentage of variation observed at a 5 cm distance between the WO and NO conditions was evaluated across a selected set of parameters. These parameters included the series inductance (L_s), impedance (Z), parallel resistance (R_p), and series resistance (R_s) of the inductor. As an example, the percentage of variation for R_p is given in Equation (2). This assessment provides a clearer insight into the frequencies that effectively represent different levels of corrosion between 0% and 100% corrosion with a better resolution.

$$\Delta R_p(\%) = \frac{R_p^{\mathrm{NO}} - R_p^{\mathrm{WO}}}{R_p^{\mathrm{NO}}} \times 100 \quad (2)$$

Here, R_p^{NO} is the no object (NO) condition and R_p^{WO} is the with object (WO) condition, where the non-corroded rebar is located in proximity to the inductor at a fixed distance. Taking into account the uncertainties of the steel bar's slight orientation that could occur when placing the bar in the setup, several measurements were performed for the same steel bar at different times to assess the percentage of variation in the parameters.

Given the results in Figure 5, only the parameters R_p and R_s (which are closely related and provide the same information) show a significant percentage of variation.

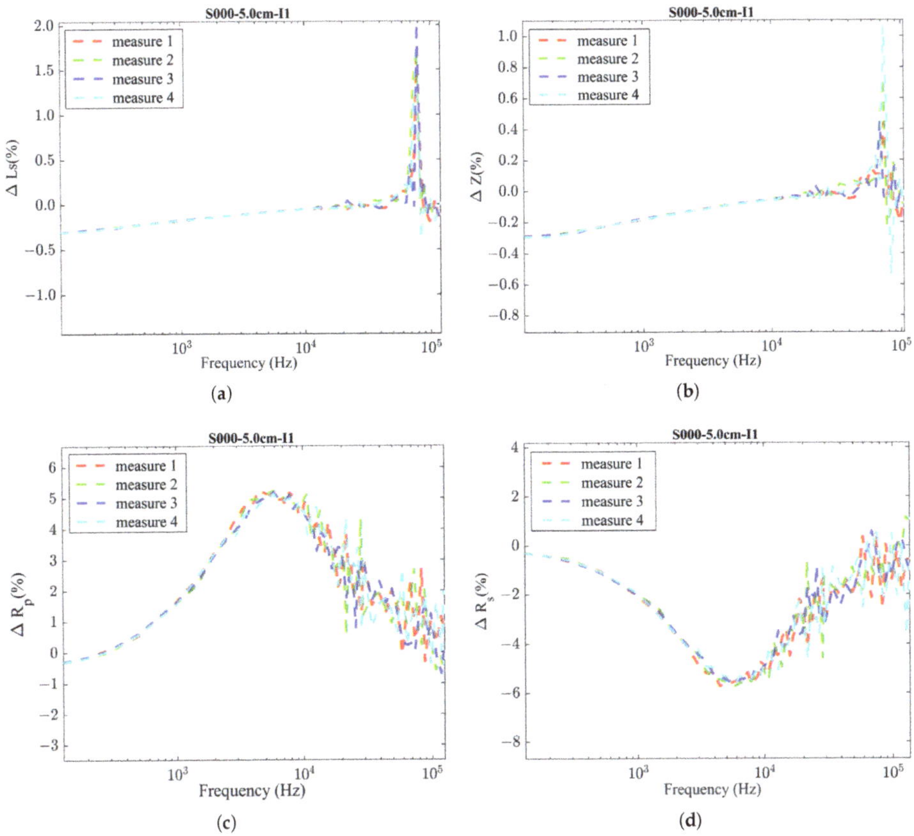

Figure 5. Inductor 1 testing results using LCR meter: % variations in the inductor parameters with a frequency sweep considering WO and NO conditions. (**a**) % variation in inductance (L_s); (**b**) % variation in impedance (Z); (**c**) % variation in parallel resistance (R_p); (**d**) % variation in series resistance (R_s).

On the other hand, L_s and Z did not seem to be suitable for the analysis. Therefore, we chose the parameter R_p as a suitable parameter to interpret the different corrosion levels. This analysis was conducted for both Inductor 1 and Inductor 2, and a comparison between the percentages of variation in R_p between the two inductors is illustrated in Figure 6. The comparative analysis confirmed that both inductors demonstrated a similar percentage of variation in R_p under the WO and NO conditions in a specific frequency range. These favorable frequency bands were identified for each inductor as 2–30 kHz for Inductor 1 and 90–700 kHz for Inductor 2. Even though we observed similar variations for both inductors, they are given as relative values compared to other tested frequencies. The actual quantified values provided by these two inductors when formed as LC probes for WO and NO conditions might present a different scenario. It is important to mention that we expect to have a considerably larger difference in R_p under the WO and NO conditions, representing 0% and 100% corrosion, as there should be a sufficient resolution to represent intermediate corrosion levels.

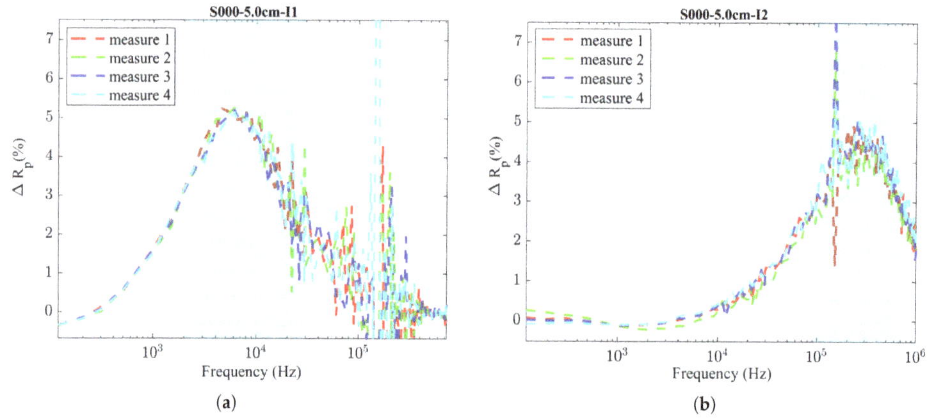

Figure 6. Comparison of percentage of variation in R_p ($\Delta R_p\%$). (**a**) Inductor 1; (**b**) Inductor 2.

Upon identifying the favorable frequency range, as shown in Figure 6, we proceeded to select the appropriate resonance frequency for the LC resonator. This involved matching each inductor with a compatible capacitor to construct an LC probe, ensuring that the probe's oscillation frequency fell within the chosen frequency range for each inductor, as well as the availability of standard capacitance components. The selected capacitor value for each LC probe was 2.7 nF for Inductor 1 and 22 nF for Inductor 2. While sweeping the frequency, the resonant frequency will correspond to the point where the impedance reaches a minimum for a series LC circuit or a maximum for a parallel LC circuit. Accordingly, the oscillating frequencies of each LC probe obtained experimentally were 24 kHz and ~180 kHz , respectively. These resonance frequency values may deviate slightly from the theoretically calculated values due to the component tolerances, parasitic capacitance, and environmental conditions during testing.

The corresponding LC probes were then tested using the LDC1000EVM device to evaluate the digital output. This evaluation aimed to determine the effectiveness of the LC probes in detecting various corrosion levels, which will be discussed in the following section.

4.2. Inductance-to-Digital Converter: LDC1000EVM

In our final sensor solution, it was decided to integrate the LC probe with an inductance-to-digital converter to obtain its digital output value. For this purpose, we utilized the LDC100x series from Texas Instruments, which incorporates inductive sensing technology, alongside the selected LC probes.

During the sensor design phase, the LC probes were tested using the LDC1000EVM evaluation module by Texas Instruments (Dallas, TX, USA), which incorporates the LDC1000 IC [20]. The LDC1000EVM device is an inductance-to-digital converter designed for parallel resistance (R_p) and inductance (L) measurements. The LDC1000EVM facilitates the evaluation of the capabilities of the LDC1000 IC and the validation of sensing circuits depending on the sensing application.

An inductor can be modeled as discussed in Section 3. Figure 7 illustrates how the connection between the LC probe and the LDC1000EVM has been formed.

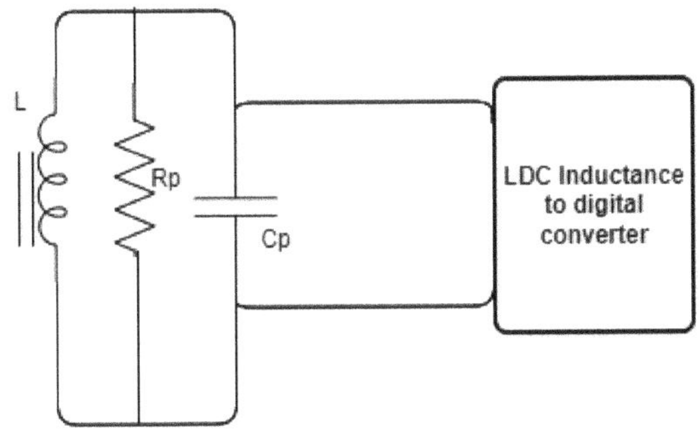

Figure 7. Equivalent resistance of RS in parallel with LC probe.

As a conductive object moves closer to or away from the coil, it changes the characteristics of the coil, which can be observed as a variation in the inductance (L) and in the series resistor (R_s). Since these variations are small in Rs but significant when converted into the parallel resistance (R_p), changing from a few kilo-ohms to hundreds of kilo-ohms, the LDC1000EVM measures the equivalent parallel resonance impedance (R_p) during the measurement.

The LDC1000 supports a wide range of LC combinations, with oscillation frequencies ranging from 5 kHz to 5 MHz and R_p ranging from 798 Ω to 3.93 MΩ. The range of R_P is typically much smaller than the maximum input range supported by the LDC1000. To enhance the resolution within the target sensing range, the LDC1000 features programmable input range capabilities, adjustable through the Rp_{min} and Rp_{max} registers. Consequently, it is very important to configure the R_p range appropriately by selecting suitable Rp_{min} and Rp_{max} limits before initiating the measurements, to yield meaningful results. The LDC1000EVM facilitates this process by allowing the user to set these parameters for the R_p limits through a graphical user interface that accompanies the device.

In the LDC1000, the resonance impedance is calculated from the digital output code (Y), as shown in Equation (3).

$$R_p = \frac{(Rp_{max} \times Rp_{min})}{(Rp_{min} \times (1 - Y) + Rp_{max} \times Y)} \quad (3)$$

There are 16 bits allocated for the R_p parameter. Considering the change in R_p with respect to Y (proximity), we obtain

$$\frac{dRp}{dY} = -\frac{(Rp_{max} - Rp_{min}) \times Rp^2}{2^{15}(Rp_{max} \times Rp_{min})} \quad (4)$$

If ΔRp is the change in R_p for 1 bit,

$$\Delta Rp = \frac{dRp}{dY} \times \Delta Y$$

where ΔY is the change in Y for 1 bit.

Thus, to obtain the number of bits to represent R_p, we denote this value as nb:

$$\frac{(Rp_{max} - Rp_{min})}{\Delta Rp} = \frac{-2^{15} \times Rp_{max} \times Rp_{min}}{Rp^2} \quad (5)$$

$$nb = log_2(\frac{2^{15} \times Rp_{max} \times Rp_{min}}{Rp^2}) = 15 + log_2(\frac{Rp_{max} \times Rp_{min}}{Rp^2}) \quad (6)$$

Reducing the error of R_p due to the noise of the measurement,

$$log_2(\frac{2^{15} \times Rp_{max} \times Rp_{min}}{Rp^2}) = 15 + log_2(\frac{Rp_{max} \times Rp_{min}}{Rp^2}) - log_2(\frac{Rp_{max} - Rp_{min}}{\Delta Rp_{corrosion}}) \quad (7)$$

Here, $\Delta R_{p_{\text{corrosion}}}$ is the value obtained from $(Rp_{\text{NO}} - Rp_{\text{WO}})$.

However, if $\Delta Rp_{corrosion} / 2^{nb} < Rp_{\text{standard deviation (WO)}}$, the measurement noise dominates the resolution.

In this case, the number of bits to represent the R_p is given by

$$\text{Number of bits to represent } R_p(nb) = log_2(\frac{\Delta Rp_{corrosion}}{2} \times \frac{1}{Rp_{\text{standard deviation (WO)}}}) \quad (8)$$

LC Probe Testing with LDC1000EVM

The LC probes were tested using the LDC1000EVM under the same two conditions, WO and NO, using the same experimental setup (see Figure 3).

During a measurement event, a set of discrete data points is obtained, which may show small variations and sometimes include outliers, as can be seen in Figure 8a,c. To address these irregularities, several filtering steps are implemented before determining the final R_p value. Outliers are managed using median filtering. Outliers in a dataset of 200 points per measurement were addressed using a moving median filter with a window size of 51. This method involves sliding a window across the dataset and calculating the median value for the data within each window. Each data point is then compared to its corresponding moving median. Those deviating significantly from the moving median (calculating the mean absolute deviation (MAD) within the window), beyond a certain threshold (3 $\times$ MAD), are identified as outliers. The detected outliers are then replaced using linear interpolation to ensure a more consistent and accurate representation. Subsequently, the median of this filtered dataset is selected as the final R_p value for this particular measurement event.

The results obtained from the LDC1000EVM reveal a greater difference between the WO and NO conditions for Inductor 1, approximately 1.5 kΩ, compared to Inductor 2, which exhibits a smaller difference of around 100 Ω. These results are illustrated in Figure 8 and were achieved by selecting the best suitable Rp_{min} and Rp_{max} limits chosen within the LDC1000EVM for each inductor. The selection of the minimum and maximum limits for the LC probe was performed by systematically testing a series of narrow limit bands close to each other. This was performed to ensure that the measured R_p value would be adequately bound by these limits, yet not excessively distant from either threshold, thereby optimizing the scaling resolution. The optimal band was identified through comparative evaluations, with a focus on achieving the best performance outcomes from the LDC1000EVM.

The R_p value is affected by the distance between the sensing coil and the conductive target. As the distance increases between the coil and the conductive object, R_p typically increases due to the decrease in eddy current losses [21]. Therefore, we can observe that R_p increases in the NO condition compared to the WO condition in the results obtained for the resonators. Furthermore, the resonance frequency of the LC resonators has also affected the R_p value in our results. A higher resonance frequency typically results in weaker eddy currents as the energy transfer at these frequencies is less efficient, particularly when there is an increased distance between the inductor and the conductive object. Therefore, considering the resonance frequency of each LC resonator, we observe differences in the R_p value range: the LC resonator with Inductor 1 at 2.7 nF exhibits an R_p value of around 72 kΩ, while the LC resonator with Inductor 2 at 22 nF has an R_p value of 3.6 kΩ.

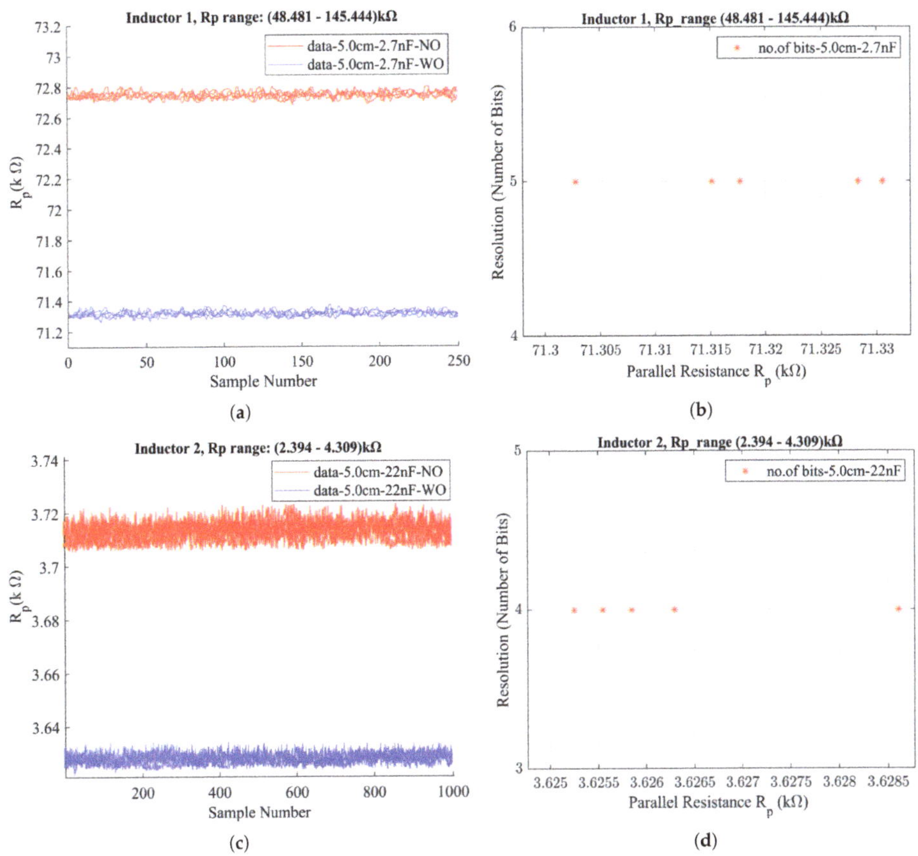

Figure 8. Results from LDC1000EVM for LC probes under WO (S000) and NO conditions. (**a**) R_p value for LC probe with Inductor 1 with 2.7 nF; (**b**) No. of bits to represent corrosion for LC probe with Inductor 1 with 2.7 nF (**c**) R_p value for LC probe with Inductor 2 with 22 nF; (**d**) No. of bits to represent corrosion for LC probe with Inductor 2 with 22 nF.

From the test results obtained for the inductors, it can be observed that better results are achieved (a larger difference in the R_p value between the NO and WO conditions) with Inductor 1, in which the inductance is in the mH range. In theory, as the inductance L increases, the resonance frequency of the inductor decreases, as in the relation shown in Equation (1); this is consistent with our experimental observations.

Moreover, it has been observed that, for both inductors, the following condition holds true:

$$\frac{\Delta R_{p_{\text{corrosion}}}}{2^{n_b}} < R_{p_{\text{standard deviation (WO)}}} \tag{9}$$

This implies that we will lose some bits from the allocated total number of bits for R_p due to measurement noise (standard deviation), as represented in Equation (8).

Considering the testing results from both the LCR meter and LDC1000EVM, the following conclusions have been drawn.

1. The parameters that effectively differentiate the corrosion levels in the inductors are only R_p, R_s.

2. Inductors with lower inductance (in the μH range) tend to exhibit smaller differences in R_p between the WO and NO conditions, indicating potentially limited sensitivity to corrosion levels.
3. Conversely, inductors with higher inductance (in the mH range) demonstrate comparatively larger differences in R_p between the WO and NO conditions, suggesting enhanced sensitivity to corrosion levels and thus being more suitable for corrosion detection applications.
4. Selecting higher inductance for the LC probe results in a lower oscillation frequency, as evidenced by the performance of Inductor 1, which oscillated at around 24 kHz; this indicates the potential for enhanced inspection depths.

Following the above conclusions, an ad hoc inductor has been designed. This new inductor maintains similar characteristics to Inductor 1 but specifically features a reduced weight and size.

4.3. Ad Hoc Inductor Design Based on Experimental Results

The characteristics of the ad hoc designed inductor are tabulated in Table 1. The eddy current coil/inductor chosen for this purpose is an air-cored circular coil that is commonly used in eddy current testing applications.

Table 1. Ad hoc inductor characteristics.

Characteristic	Value
Inductor inner diameter	10 cm
Wire diameter	0.3 mm
Number of turns	110
Self-resonance frequency (tested using LCR meter HIOKI IM3536)	350 kHz
Inductance in the 35–65 kHz frequency range (tested using LCR meter HIOKI IM3536)	3 mH

The ad hoc designed inductor is named Inductor 3 hereafter. Inductor 3 was also tested under the same procedure: the initial phase involved characterizing Inductor 3 using an LCR meter, followed by assessments using the LDC1000EVM. Additionally, the frequency range of Inductor 3 was tested at a distance of 6 cm, considering that the concrete thickness can vary between 5 and 6 cm. For both scenarios, 35–65 kHz was selected as the suitable operational frequency range of the LC probe (see Figure 9).

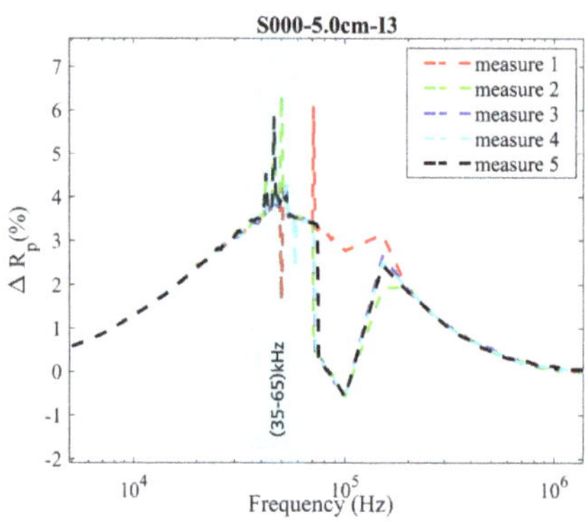

(**a**) At 5 cm distance

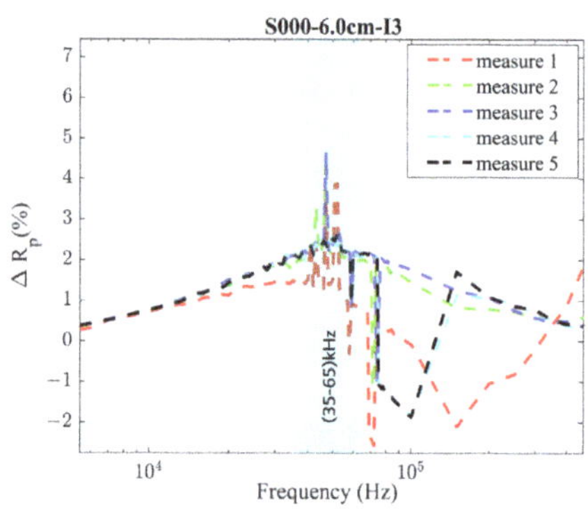

(**b**) At 6 cm distance

Figure 9. $\Delta R_p(\%)$ study using Inductor 3 with a frequency sweep at a distance of (**a**) 5 cm and (**b**) 6 cm (measurements from Hioki IM3536 LCR meter).

As the resonance frequency falls within this range, three capacitor values of 6.2 nF, 3.6 nF, and 2.2 nF were tested for the LC probe, each paired with Inductor 3. The experimentally obtained resonance frequencies for each pairing were 38 kHz, 48.5 kHz, and 62 kHz, respectively. Then, each LC probe configuration was assessed using the LDC1000EVM by Texas Instruments (Dallas, TX, USA). After thorough testing, the 2.2 nF capacitor was identified as the optimal selection for the LC probe. The performance of the LC probe equipped with the 2.2 nF capacitor and Inductor 3 is illustrated in Figure 10.

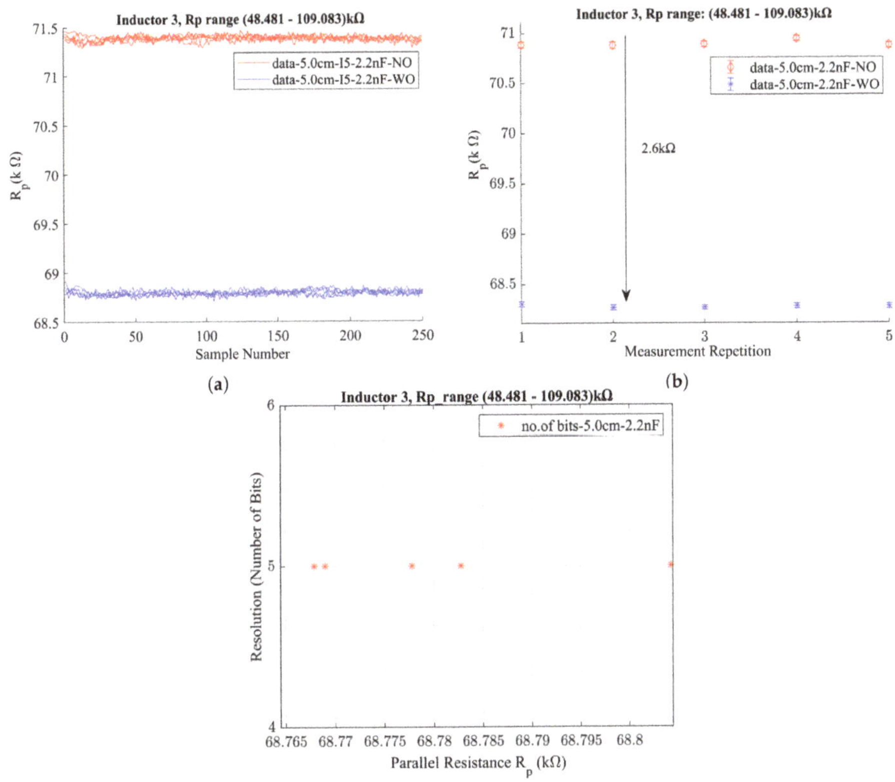

(c)

Figure 10. LC probe (with Inductor 3 and 2.2 nF capacitor) testing with sample S000 using LDC1000EVM. (**a**) R_p variation. (**b**) Final R_p value estimated for 5 measurements. (**c**) Number of bits to represent corrosion condition.

According to the results obtained from the LDC1000EVM, the LC probe with Inductor 3 and 2.2 nF capacitor demonstrates improved performance compared to the LC probe we tested with Inductor 1 and 2.7 nF capacitor. The difference of R_p value could obtain around 2.6 kΩ and 5 bits to represent the corrosion condition.

In an effort to further refine the LC probe's performance, a ferrite plate with a thickness of 5 mm was positioned on top of Inductor 3. This adjustment aimed to direct the magnetic field lines downward, thereby improving the focus and sensitivity of the probe. With the new modification, we achieved an R_p difference of approximately 7 kΩ between the WO and NO conditions under the same distance conditions as before (with the sensor-to-rebar distance of 5 cm) and with an average of 6 bits to represent the corrosion condition. Based on these promising results, we finalized our LC probe design by incorporating a 5 mm ferrite plate in addition to the LC probe—Inductor 3—and the 2.2 nF capacitor, as shown in Figure 11.

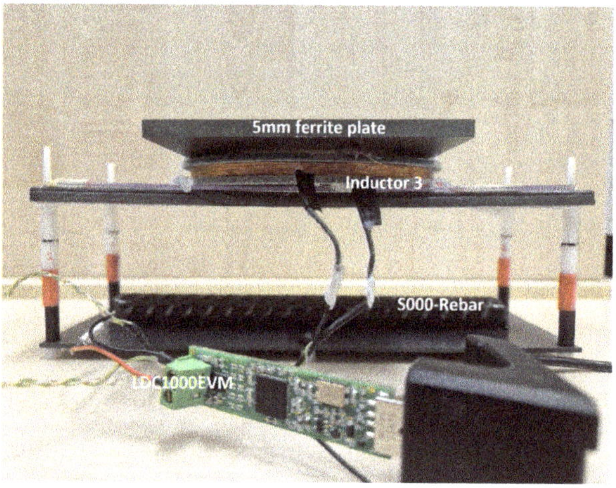

Figure 11. EC sensor with final LC probe and LDC1000EVM placed on the test setup.

Subsequently, the EC sensor, which combines the LC probe with an LDC1000 inductance-to-digital converter, will be integrated with the necessary front-end electronics, including wireless communication capabilities, to develop a complete EC sensor node. This sensor node will then be ready for field testing integration.

The final version of the LC probe was subjected to an experiment to measure the variation in the parallel resistance, R_p, with temperature fluctuations. This investigation aimed to quantify the necessary adjustments to R_p that account for temperature variations, ensuring that the probe provides consistent and reliable readings that can be attributed solely to the corrosion levels, rather than being affected by temperature-induced changes. More details of the experiment are discussed in the following section.

4.4. *Temperature Experiment*

The test setup with the inductor and the 5 mm ferrite plate was placed inside a thermal chamber (VT 4011 from Vötsch Industrietechniks) (refer to Figure 12).

During the temperature experiment, the capacitor in the LC probe and other electronics were placed outside the chamber. The variation in the parallel resistance (R_p), as a function of the temperature, was observed across a range extending from 0 °C to 45 °C. The measurements were carried out at every 5 °C interval within this range, allowing the temperature to stabilize before the measurements. The experiment was conducted under the with object (WO) condition, which we assumed to be closer to the real scenario, with the hypothesis being that the temperature variations affecting both the rebar and the sensor would be similar. However, this assumption has its limitations, as it may not fully reflect the real temperature conditions. Given that, in practical applications, the rebars are embedded within a concrete structure, this can lead to temperature gradients or delays in temperature equilibration between the rebar and the sensor placed externally.

The first part of the experiment was conducted with the sensor placed 5 cm away from the rebar. This experiment was conducted bidirectionally, with temperature variations being tested both from 0 °C to 45 °C and in reverse, from 45 °C to 0 °C. This approach helped to account for any hysteresis effects in the R_p response due to temperature changes, ensuring that the R_p value's dependence on the temperature was consistent and reproducible regardless of the direction of the temperature change. The relationship between R_p and the temperature was nonlinear, best described by a quadratic model. The results indicated similar fitting in both directions, with R_p decreasing as the temperature increased. Based on the temperature fitting data obtained for temperature compensation in the temperature range of 20–25 °C (considered the normal laboratory ambient temperature range), it is estimated that an approximately 155 Ω variation in parallel resistance (R_p) can be expected for every 1 °C change in temperature.

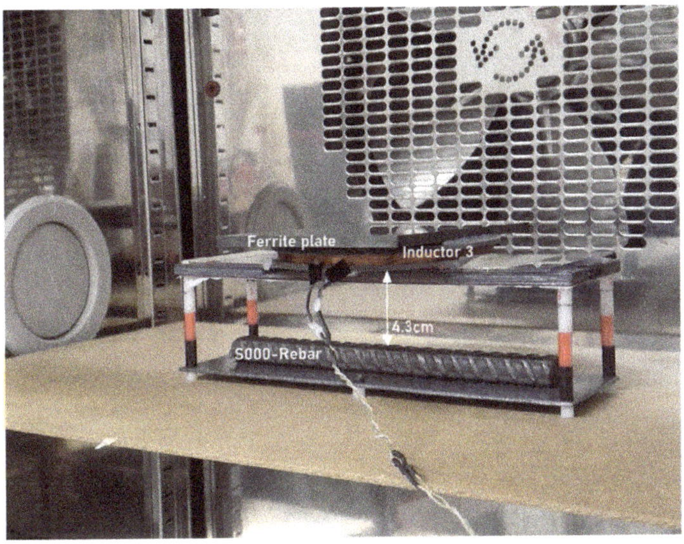

Figure 12. Experimental setup inside the thermal chamber; study of R_p value under WO (S000) condition.

Following this, the study was extended to a reduced distance of 4.3 cm between the sensor and the rebar. Comparing both distances, the results showed that the coefficients for the temperature fitting were very similar (see Figure 13), demonstrating a consistent thermal response regardless of the distance.

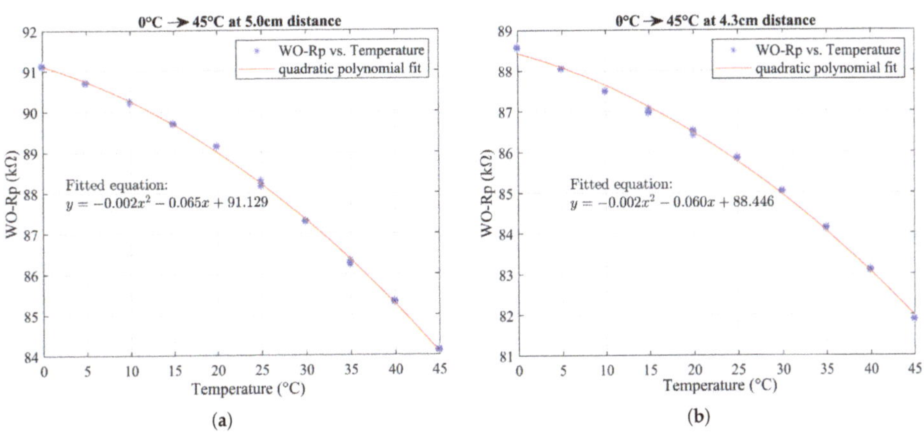

Figure 13. Comparison of R_p response to temperature variations (**a**) at 5 cm distance; (**b**) at 4.3 cm distance.

Considering the quadratic fitting equation $y = ax^2 + bx + c$, where a, b, and c are the coefficients of fitting, Table 2 presents the fitting coefficients obtained from each group of temperature experiments.

Table 2. Quadratic fitting coefficients obtained in temperature experiment.

Control Experiment	Distance (cm)	Temp. Cycle (°C)	Coeff. *a*	Coeff. *b*
1	5	0–45	−0.002	−0.065
2	5	45–0	−0.002	−0.067
3	4.3	0–45	−0.002	−0.060

The experimental results demonstrate that the sensor–rebar distance is not a significant factor affecting the thermal relationship with R_p, which simplifies the temperature compensation since similar behavior is observed even when variations in sensor placement occur.

5. Concrete Experiment with EC Sensor

Our EC sensor's performance was tested under conditions closely simulating reinforced corrosion detection scenarios. An experiment was designed to assess the rebar conditions within a high-performance concrete (HPC) block measuring 15 × 15 × 15 cm. A hole was created inside the block to accommodate the rebar. The experiment aimed to test the sensor's performance across different corrosion levels by mechanically degrading the rebar to simulate corrosion. The rebar sample (S001) used in the experiment was 50 cm long. The core, representing the cylindrical section of the bar, had a diameter of approximately 11.5 mm. Additionally, the corrugations on the surface of the rebar, which varied in height, measured between 0.5 mm and 1 mm.

At first, measurements were taken with the rebar at its initial, non-degraded state. Subsequently, uniform material degradation was applied along the rebar to simulate various corrosion levels. During the first degradation step, the material was selectively removed to achieve an approximately 3% reduction in the diameter of the cylindrical section of the rebar, resulting in an additional amount of material being removed due to the presence of surface corrugation.

The testing setup, shown in Figure 14c,d, involved placing the rebar inside the concrete block and positioning the sensor on the outer surface during the measurements.

The measured distance between the sensor and the top of the rebar was 4.3 cm. Throughout the experiment, efforts were made to keep the bar orientation unchanged. In real conditions, the rebar remains fixed. To closely replicate real-world conditions and ensure the consistency and relevance of our methodology, we maintained the same orientation of the rebar.

The main objective of this experiment was to establish a definitive correlation between the fluctuations in parallel resistance (R_p), as detected by our EC sensor, and the weight loss of the rebar, which is indicative of varying levels of corrosion. To ensure that the variations in R_p were attributed solely to material loss, the rebar was kept at a consistent elevation relative to the sensor. As we induced uniform degradation along the length of the rebar, a relevant shift in the mid-axis level was expected due to the material loss occurring uniformly around the rebar once it was placed within the concrete. However, we controlled this by keeping the central axis at the same height throughout the measurement process after each degradation step. In actuality, in the case of uniform corrosion, the central axis of the rebar generally remains in the same position during the corrosion process, because the corrosion products, which can include rust and other compounds, often fill the spaces left by the corroded metal. This means that although the rebar is losing its original material to corrosion, the volume may appear to be conserved due to the accumulation of these products. By simulating this aspect in our experimental setup, we aimed to measure the changes in R_p that would directly correspond to the actual loss of material, as would be the case in a natural corrosion process.

(**a**)

(**b**)

(**c**)

(**d**)

Figure 14. The rebar (S001) measurements with the mechanical degradation process. (**a**) Sample: S001. No degradation; (**b**) Sample: S001 After first degradation; (**c**) Concrete test before any degradation; (**d**) Concrete test after first degradation.

Results and Discussion

The mechanical degradation simulation, aimed at replicating reinforcing bar corrosion, yielded results across four distinct stages of degradation. The first was 0% corrosion, where we measured the non-corroded S001 bar without any degradation. This was followed by the first degradation stage, where the S001 bar was degraded with 10% weight loss from its initial weight. Afterwards, the second degradation stage was assessed, with the 3% weight loss of the bar; finally, 100% corrosion was considered, where we measured it without the presence of the rebar to simulate complete corrosion to represent 100% corrosion.

After the first degradation, the results showed a difference of approximately 1.2 kΩ in the R_p value compared to the non-degraded state. This corresponds to an average of 11% weight loss and a 14% change in the R_p value. Since the difference in R_p between the two states is quite high, there is potential to detect lower percentages of degradation than the observed 11%. Therefore, the second degradation was performed as we lost approximately 3% of the material, with a reduction in diameter of around 0.17 mm.

As the results shown in Figure 15, we obtained a R_p range of 8.45 kΩ between 0% and 100% corrosion, providing a substantial range to represent various degrees of corrosion.

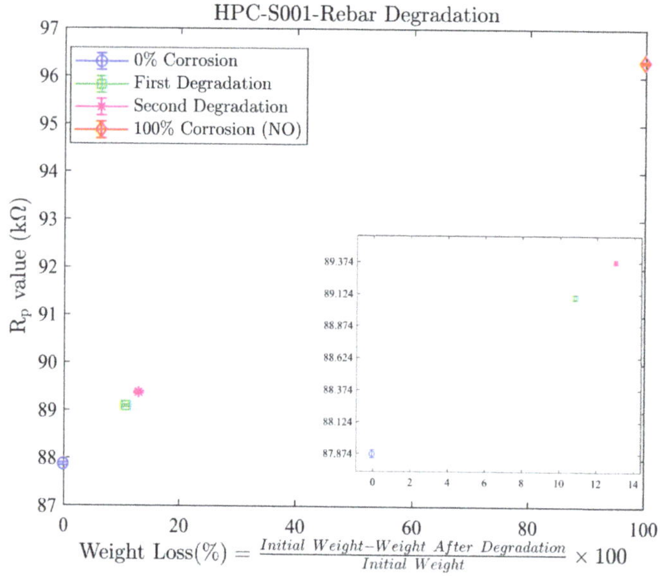

Figure 15. R_p value vs. corresponding weight during the corrosion experiment with rebar S001.

After the second degradation step, there was a 275 Ω increase in the R_p value compared to the first degradation, corresponding to 3% weight loss in the rebar. This means that the sensor is capable of detecting 3% weight loss, accounting for material loss around the whole rebar, resulting in less than a 0.2 mm (approximately 0.17 mm) reduction in diameter.

The amount of the R_p increase observed during the second degradation test is substantial enough to distinguish another level of corrosion without being affected by the standard deviation of the measurements. Typically, after handling outliers, the standard deviation of the measurements normally ranges between 20 Ω and 60 Ω, as shown in Figure 16.

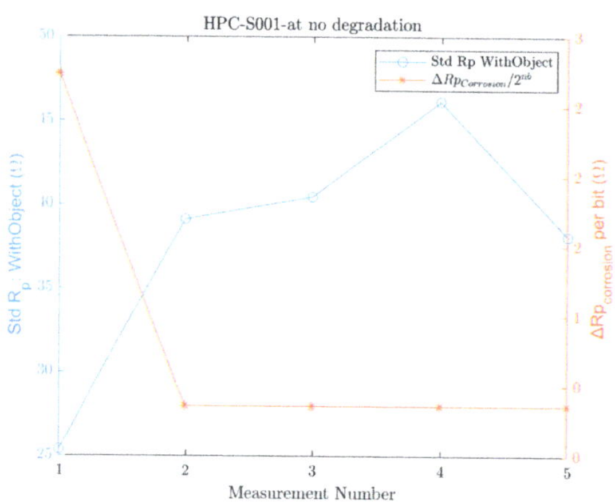

Figure 16. Change in R_p for 1 bit of the register allocated for R_p and the standard deviation of 5 consecutive measurements performed for non-corroded S001.

Figure 16 presents the standard deviation of the measurements conducted at 0% corrosion (no-degradation) during the concrete test, which can be generalized to the performance of the final version of the EC sensor.

Furthermore, a comparison between the standard deviation of the measurements and the ΔR_p that is the R_p value represented per 1 bit under the test conditions is presented, indicating that, typically, the standard deviation of the measurements influences the resolution, as discussed in Section 4.2. Hence, Equation (8) shows the relationship that governs the number of bits to represent the measured R_p values.

Considering the impact of the temperature on R_p, our experimental results showed that, for each 1 °C deviation within the 20–25 °C range, the R_p values need to be corrected by 155 Ω. Although the laboratory conditions were relatively stable, with minor temperature variations, in real-world scenarios, where the environmental conditions are far more variable, temperature-compensated data are essential to account for the influence of temperature fluctuations on the R_p measurements for the relative assessment of corrosion conditions.

The uniform mechanical degradation cycles employed in our experiments are indeed a simplification of the natural corrosion process, primarily aimed at testing our sensor's performance under the accelerated wear that materials might experience in certain offshore conditions. This approach, while simulating uniform corrosion along the rebar (S001) at different stages, does not fully capture the complex chemical interactions and environmental factors inherent in natural corrosion. Hence, any effect on R_P influenced by a variety of environmental conditions, such as humidity and the presence of corrosive agents, has not been accounted for. The synergistic effects of these parameters and their fluctuations over long periods, as they occur in nature, present a challenge for the laboratory simulation. Additionally, the rate of degradation in our simulations may not reflect the typically slower pace of natural corrosion, as the material degradation limitations and time are often condensed in a laboratory setting for feasibility.

6. Conclusions and Future Works

This paper details the development and application of a single-frequency eddy current (EC) sensor employing an LC resonator for the non-destructive evaluation of corrosion levels within concrete-embedded reinforcing bars. The sensor's design and experimental validation process utilized a comprehensive methodology to study the optimal LC resonator characteristics, aiming to develop a system capable of detecting changes in the rebar condition from a distance of 5–6 cm outside the concrete structure, as per the application requirements. Following the experimental methodology, the EC sensing design integrates an LC probe with a circular air-cored inductor of approximately 3 mH inductance and a 2.2 nF capacitor connected in parallel, along with a 5 mm ferrite plate placed on top of the inductor.

The temperature effect on the sensor's output (R_p) was evaluated using a temperature experiment under conditions in which the eddy current coil was positioned facing the S000 rebar at a 5 cm distance. The change in R_p per 1 °C in the temperature range of 20–25 °C was recorded as 155 Ω. The temperature study results showed that the sensor–rebar distance was not a significant factor affecting the thermal relationship with R_p.

Through the systematic simulation of uniform corrosion via mechanical degradation cycles, we have demonstrated the sensor's capability to accurately quantify material loss with a resolution of approximately 0.17 mm.

While the results are promising under laboratory conditions, it is acknowledged that the mechanical degradation approach employed does not fully replicate the complex nature of real-world corrosion. Therefore, further research is required to incorporate environmental factors to enhance the sensor's practicality under real offshore operating conditions. In future work, we aim to enhance the validity of our findings by testing the sensor's performance under conditions that more accurately reflect real-world applications. To achieve this, the sensor will be integrated into a concrete block with real rebar conditioning (a real reinforced structure block) and tested under offshore environmental conditions. The next step will help us to identify any potential performance differences between laboratory and field conditions and refine the sensor design accordingly.

Author Contributions: Conceptualization, A.C. and A.I.; Formal analysis, A.I. and U.C.T.; Funding acquisition, A.C.; Investigation, A.C., A.I. and U.C.T.; Methodology, A.C., A.I. and U.C.T.; Project administration, A.C.; Software, A.I. and U.C.T.; Supervision, A.C. and A.I.; Validation, A.C., A.I., A.M.C. and U.C.T.; Writing—original draft, U.C.T.; Writing—review and editing, A.C., A.I. and A.M.C. All authors have read and agreed to the published version of the manuscript.

Funding: The research work is part of the SUREWAVE project, Funded by the European Union with GA No. 101083342. Views and opinions expressed are however those of the author(s) only and do not necessarily reflect those of the European Union. Neither the European Union nor the granting authority can be held responsible for them.

Institutional Review Board Statement: Not applicable.

Informed Consent Statement: Not applicable.

Data Availability Statement: Data are contained within the article.

Acknowledgments: This work has been possible thanks to the cooperation of CEIT with all of the SUREWAVE partners, especially, in this case, with ACCIONA, which was responsible for producing the samples, and the MEC group of CEIT, which provided us with the necessary support to degrade the rebars mechanically during the experiment.

Conflicts of Interest: The authors declare no conflicts of interest.

Abbreviations

The following abbreviations are used in this manuscript:

AC	Alternating Current
ECT	Eddy Current Testing
EC	Eddy Current
EMF	Electromotive Force
HCP	Half-Cell Potential
HPC	High-Performance Concrete
LPR	Linear Polarization Resistance
MAD	Median Absolute Deviation
NDT	Non-Destructive Testing
NO	No Object
OCP	Open Circuit Potential
PV	Photovoltaic
R_p	Parallel Resistance
Rebar	Reinforcing Bar
SRF	Self-Resonance Frequency
SP	Surface Potential
UT	Ultrasonic Testing
WO	With Object

References

1. Qiao, G.; Ou, J. Corrosion monitoring of reinforcing steel in cement mortar by EIS and ENA. *Electrochim. Acta* **2007**, *52*, 8008–8019. [CrossRef]
2. Zhao, Y.; Karimi, A.R.; Wong, H.S.; Hu, B.; Buenfeld, N.R.; Jin, W. Comparison of uniform and non-uniform corrosion induced damage in reinforced concrete based on a Gaussian description of the corrosion layer. *Corros. Sci.* **2011**, *53*, 2803–2814. [CrossRef]
3. Rodrigues, R.; Gaboreau, S.; Gance, J.; Ignatiadis, I.; Betelu, S. Reinforced concrete structures: A review of corrosion mechanisms and advances in electrical methods for corrosion monitoring. *Constr. Build. Mater.* **2021**, *269*, 121240. [CrossRef]
4. Naidu Gopu, G.; Joseph, S.A. Corrosion Behavior of Fiber-Reinforced Concrete—A Review. *Fibers* **2022**, *10*, 38. [CrossRef]
5. Brijder, R.; Hagen, C.H.; Cortés, A.; Irizar, A.; Thibbotuwa, U.C.; Helsen, S.; Vásquez, S.; Ompusunggu, A.P. Review of corrosion monitoring and prognostics in offshore wind turbine structures: Current status and feasible approaches. *Front. Energy Res.* **2022**, *10*, 991343. [CrossRef]

6. Song, H.W.; Saraswathy, V. Corrosion monitoring of reinforced concrete structures-A. *Int. J. Electrochem. Sci.* **2007**, *2*, 1–28. [CrossRef]
7. Shevtsov, D.; Cao, N.L.; Nguyen, V.C.; Nong, Q.Q.; Le, H.Q.; Nguyen, D.A.; Zartsyn, I.; Kozaderov, O. Progress in Sensors for Monitoring Reinforcement Corrosion in Reinforced Concrete Structures—A Review. *Sensors* **2022**, *22*, 3421. [CrossRef] [PubMed]
8. AbdAlla, A.N.; Faraj, M.A.; Samsuri, F.; Rifai, D.; Ali, K.; Al-Douri, Y. Challenges in improving the performance of eddy current testing. *Meas. Control* **2019**, *52*, 46–64. [CrossRef]
9. Lu, M.; Meng, X.; Huang, R.; Chen, L.; Peyton, A.; Yin, W.; Qu, Z. Thickness measurement of circular metallic film using single-frequency eddy current sensor. *NDT & E Int.* **2021**, *119*, 102420.
10. Yin, W.; Peyton, A. Thickness measurement of non-magnetic plates using multi-frequency eddy current sensors. *NDT E Int.* **2007**, *40*, 43–48. [CrossRef]
11. Kumar, K.; Muralidharan, S.; Manjula, T.; Karthikeyan, M.; Palaniswamy, N. Sensor systems for corrosion monitoring in concrete structures. *Sens. Transducers* **2006**, *67*, 553–560.
12. Song, M.; Li, M.; Xiao, S.; Ren, J. Research on the influence of geometric structure parameters of eddy current testing probe on sensor resolution. *Sensors* **2023**, *23*, 6610. [CrossRef]
13. Chen, W.; Wu, D. Resistance-frequency eddy current method for electrical conductivity measurement. *Measurement* **2023**, *209*, 112501. [CrossRef]
14. Andringa, M.; Neikirk, D.; Dickerson, N.; Wood, S. Unpowered wireless corrosion sensor for steel reinforced concrete. In Proceedings of the SENSORS, 2005 IEEE, Irvine, CA, USA, 30 October–3 November 2005; 4p. [CrossRef]
15. Chen, W.; Wu, D.; Wang, X.; Wang, T. A self-frequency-conversion eddy current testing method. *Measurement* **2022**, *195*, 111129. [CrossRef]
16. Zhang, G.C.; Feng, Y.Q.; Xue, S.; Zhang, J. Experimental Research of Eddy Current Testing Technology Based on Resonant Circuit in Carbon Fiber Reinforced Polymer Testing. In Proceedings of the 2021 IEEE 12th International Conference on Mechanical and Intelligent Manufacturing Technologies (ICMIMT), Cape Town, South Africa, 13–15 May 2021; pp. 230–238. [CrossRef]
17. Texas Instruments. Application Report: Sensor Design for Inductive Sensing Applications Using LDC. Available online: https://www.ti.com/lit/an/snoa930c/snoa930c.pdf?ts=1718350012469 (accessed on 23 May 2024).
18. Würth Elektronik. WE-WPCC Wireless Power Array. Available online: https://www.we-online.com/components/products/datasheet/760308103145.pdf (accessed on 20 May 2024).
19. TexasInstruments. LDC1000EVM - Evaluation Module for Inductance to Digital Converter with Sample PCB Coil. Available online: https://www.ti.com/tool/LDC1000EVM (accessed on 17 May 2024).
20. Texas Instruments. LDC1000 Datasheet (PDF)—Texas Instruments. Available online: https://pdf1.alldatasheet.com/datasheet-pdf/view/836798/TI1/LDC1000.html (accessed on 23 May 2024).
21. Texas Instruments. LDC1001 Inductance-to-Digital Converter Datasheet—Texas Instruments. Available online: https://www.ti.com/lit/ds/symlink/ldc1001.pdf (accessed on 3 November 2023).

sensors

MDPI

Article

Research on Eddy Current Probes for Sensitivity Improvement in Fatigue Crack Detection of Aluminum Materials

Qing Zhang [1], Jiahuan Zheng [1], Shengping Wu [2], Yanchang Wang [3], Lijuan Li [1] and Haitao Wang [4,*]

1 College of Electrical Engineering and Control Science, Nanjing Tech University, Nanjing 211816, China; zhangqing@njtech.edu.cn (Q.Z.); 18652010963@163.com (J.Z.); ljli@njtech.edu.cn (L.L.)
2 Directly Affiliated Branch Institute, Special Equipment Safety Supervision Inspection Institute of Jiangsu Province, Nanjing 210009, China; wspcn@sina.com
3 713th Research Institution, China State Shipbuilding Corporation Limited, Zhengzhou 450052, China; 13203734566@163.com
4 College of Automation, Nanjing University of Aeronautics and Astronautics, Nanjing 211106, China
* Correspondence: htwang@nuaa.edu.cn

Abstract

Aluminum alloys under long-term service or repetitive stress are prone to small fatigue cracks (FCs) with arbitrary orientations, necessitating eddy current probes with focused magnetic fields and directional selectivity for reliable detection. This study presents a flexible printed circuit board (FPCB) probe with a double-layer planar excitation coil and a double-layer differential receiving coil. The excitation coil employs a reverse-wound design to enhance magnetic field directionality and focusing, while the differential receiving coil improves sensitivity and suppresses common-mode noise. The probe is optimized by adjusting the excitation coil overlap and the excitation–receiving coil angles to maximize eddy current concentration and detection signals. Finite element simulations and experiments confirm the system's effectiveness in detecting surface cracks of varying sizes and orientations. To further characterize these defects, two time-domain features are extracted: the peak-to-peak value (ΔP), reflecting amplitude variations associated with defect size and orientation, and the signal width (ΔW), primarily correlated with defect angle. However, substantial overlap in their value ranges for defects with different parameters means that these features alone cannot identify which specific parameter has changed, making prior defect classification using a Transformer-based approach necessary for accurate quantitative analysis. The proposed method demonstrates reliable performance and clear interpretability for defect evaluation in aluminum components.

Keywords: fatigue cracks; eddy current probe; magnetic field focusing; directional selectivity

1. Introduction

Aluminum alloys are commonly used in aerospace, transportation, and energy equipment owing to their light weight, corrosion resistance, and high electrical conductivity [1,2]. However, long-term operation or repetitive stress makes some areas of aluminum alloys, particularly joints, welded regions, and stressed components, susceptible to fatigue cracking (FC) [3,4]. FC grows from a small scale to a significant one, proliferating rapidly under increased stress, which may lead to material failure [5]. Thus, the detection of FC is essential for the integrity of aluminum alloy.

Eddy current testing (ECT) has been widely employed for aluminum alloy crack detection owing to its non-contact nature, rapidity, and sensitivity to surface defects [6–8]. While the orientation of the FC is uncertain, it is influenced by several factors, such as stress direction, crystal structure, and external stresses [9,10]. ECT has varied sensitivity when detecting cracks at different orientations [11,12]. It has poor sensitivity for detecting cracks that are not perpendicular to the induced eddy current caused by the probe. Moreover, early FC is narrower, requiring excellent precision of the ECT probe [13]. Therefore, it is essential to develop a probe that is highly sensitive to FC in any orientation.

Research has been carried out in the past few years, and some directional ECT probes have been developed for crack detection in any orientation. Wu et al. [14] designed a Tx-Rx coil sensor with an 8-shaped excitation coil and a circular receiver coil, demonstrating effective directionality for the detection of carbon-fiber-reinforced plastic (CFRP). Wang et al. [15] developed a probe that combines a right-angled triangular excitation coil and a TMR sensor array, which can accurately detect defects in all directions. Ge et al. [16,17] employed a rotating ECT probe that is excited by two sinusoidal waves of identical amplitude but with a 90° phase difference, to produce directional eddy currents for the identification of complicated cracks. Furthermore, some focused probes have also been proposed to enhance sensitivity. Xu et al. [18] developed a rotating focusing ECT probe by connecting four coils to form two figure-8-shaped focusing sub-probes, which are powered by two identical harmonic currents with a 90-degree phase difference. Trung et al. [19,20] developed a double semi-circular ECT probe that focuses eddy current on the surface of the specimen via a copper core. Liu et al. [21] proposed a semi-ellipsoidal eddy current probe with a magnetic flux-convergence design, incorporating a curved trapezoidal cross-section and a super-permalloy core. This configuration enhances magnetic field focusing and improves defect detection sensitivity through optimized structural parameters. Wang et al. [22] developed a Koch curve-based ECT probe that improves electromagnetic field focusing via spatial degree of freedom design, thus enhancing sensitivity to defects at different depths and orientations. However, the previously proposed probes are bulky and unsuitable for specimens with curvature. The ECT probe, constructed from flexible printed circuit board (FPCB), is then examined. Li et al. [23] introduced a probe employing an FPCB that generates directional eddy currents, enabling the detection of cracks in various orientations on complex geometric surfaces. She et al. [24] introduced a floral eddy current probe, in which the excitation coils are arranged in reverse order and symmetrically positioned. This configuration focuses the magnetic field and generates directional eddy currents through periodic distribution. Chen et al. [25] developed a fractal geometry-based FPCB probe that improves sensitivity to cracks in various orientations. To further improve its performance, they included a magnetic particle composite film into the probe, which focused the magnetic field and increased the eddy current intensity, thereby enabling more accurate detection of cracks in various orientations [26]. Trung et al. [27] designed a flexible eddy current (FEC) sensor with co-directional interleaved square coils. This compact structure enhances magnetic field focus and spatial resolution, enabling effective small crack detection at low frequencies. Ma et al. [28] presented a flexible square winding array structure that employs a double-period square excitation coil to improve directivity and focusing effects, hence augmenting the ability to detect small cracks. Compared to bulky ECT probes, only a few FPCB probes simultaneously achieve effective magnetic field focusing and the capability to detect defects across multiple orientations.

This paper proposes a focusing and directional FPCB probe aimed at improving sensitivity for the detection of FC in any orientation. The probe consists of an excitation coil and a receiving coil. The excitation coil consists of a double layer with opposing winding orientations, producing a focused diamond-shaped eddy current distribution on the surface of the aluminum alloy. The receiver coil employs a differential dual-layer configuration that enhances signal variations caused by cracks. The rest of the paper is organized as follows: Section 2 presents the design and optimization of the probe. Section 3 analyzes the probe's ability to detect defects at varying angles and compares the responses of the designed coil with the commonly used rectangular coil. Section 4 describes the setup of the experimental platform. Section 5 focuses on the analysis and evaluation of defects. Section 6 presents the conclusions.

2. Probe Design

2.1. Probe Structure

The proposed probe structure is illustrated in Figure 1a, where the receiver coil is positioned above the excitation coil. The excitation coil, depicted in Figure 1b, features a double-layer construction, with the winding orientations of the upper and bottom coils being opposite. The current of the excitation coils is illustrated in Figure 1b, with blue and red arrows denoting the current directions in the upper and bottom coils, respectively. The configuration of the excitation coil is pivotal in influencing the eddy current field on the specimen's surface. Regions with opposing current directions in the upper and bottom coils diminish eddy currents in the specimen, whereas regions with same directions enhance eddy currents. As shown in Figure 1b, "×" represents the region where eddy current is smaller, whereas "○" indicates the eddy current focus region. The distribution of eddy currents on the specimen's surface displays a diamond pattern. This not only focuses the eddy currents but also introduces a directional eddy current distribution in the specimen, which is beneficial for FC detection. Moreover, as shown in Figure 1c, two coils with the same structure but opposite winding directions are used to form the differential receiver coil, which is used to further increase the sensitivity for FC detection. The left and right halves of the probe are symmetrical to increase the detection area and improve detection efficiency.

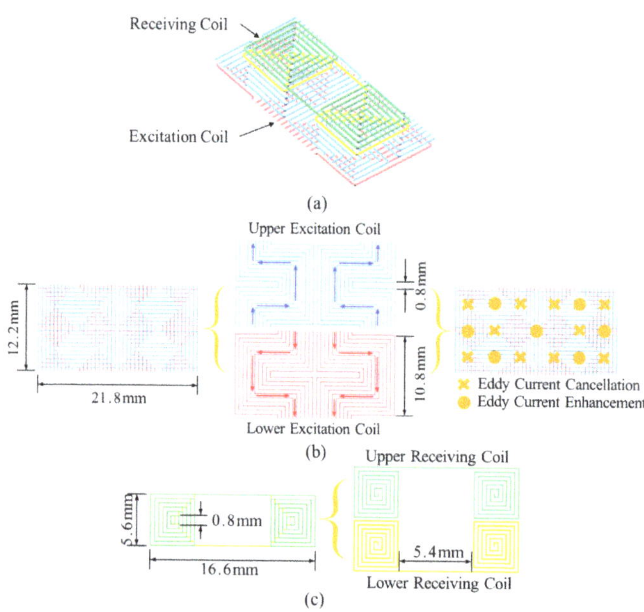

Figure 1. Proposed probe structure: (**a**) Integrated coil. (**b**) Excitation coil. (**c**) Receiving coil.

2.2. Simulation Analysis

Simulation analysis is conducted to investigate the performance of the proposed probe. The AC/DC module of COMSOL Multiphysics 6.2 is employed in the simulation analysis, and the simulation model of the proposed probe is shown in Figure 2, in which the probe is situated 0.5 mm above an aluminum plate that has an FC with 8 mm length, 0.35 mm width, and 2 mm depth. Table 1 presents the other parameters of the simulation model, including the permeability, conductivity, size of the coil, the specimen, and the air around the specimen. A Cartesian coordinate is established for description, with the crack's center position used as the origin. The crack's angle is defined as the angle formed between the crack and the x-axis. The angle of the crack seen in Figure 2 is 90°. To simulate scanning, the displacement of the probe is realized through a parametric sweep along the x-axis. The coil position is updated step by step from $x = -15$ mm to $x = 15$ mm while maintaining a constant lift-off, and the results are depicted in Figure 3.

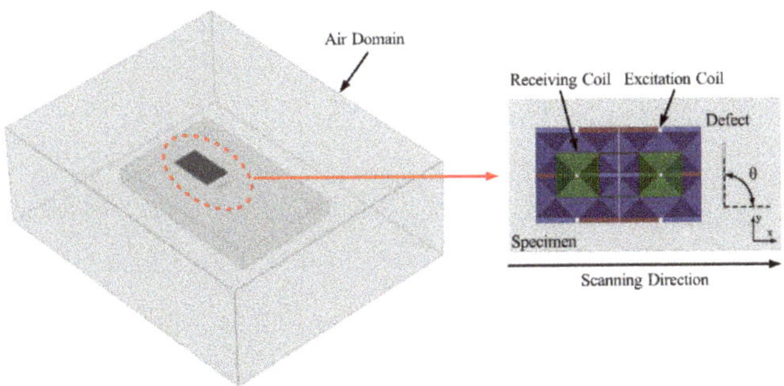

Figure 2. Finite element simulation model for defect detection analysis.

Table 1. Simulation parameters.

Parameter	Specimen	Excitation Coil	Receiving Coil	Air Domain
Materials	6061-T6	Copper	Copper	Air
Dimension (mm)	80.0/50.0/8.0	19.0/12.0/0.017	5.6/16.6/0.017	120.0/90.0/90.0
Relative permeability	1	1	1	1
Conductivity (S/m)	2.87×10^7	5.99×10^7	5.99×10^7	1

Figure 3a depicts the eddy current field on the specimen's surface induced by the proposed excitation coil at 1 MHz. The distribution of eddy currents is focused on the specimen's surface in a diamond pattern, which is consistent with the analysis presented in Section 2.1. Moreover, the signal for 90° FC is shown in Figure 3b, where the x-axis denotes the probe position and the y-axis represents amplitude. As shown in Figure 3b, the intervals $x \in (-15, -11)$ and (11, 15) mm indicate that the FC is outside the probe's detection range (which is determined by the sizes of the receiver coil), resulting in a signal amplitude of 0 V; the interval $x \in (-11, 0)$ mm signifies that the right region of the receiver coil is positioned over the FC, with the signal amplitude initially increasing and then decreasing; and at $x = -5.5$ mm, which is the center of the right receiving coil that coincides with the defect, the signal amplitude reaches its maximum value. A similar phenomenon occurs with the left receiver coil, however with inverted polarity of the amplitude, and when $x = 0$ mm, which is the receiver coil's center is positioned above the FC, its signal amplitude registers as zero. In summary, the suggested differential receiving coil is exclusively responsive to the FC and its placement, thereby enhancing the signal-to-noise ratio (SNR) and facilitating the analysis of FC.

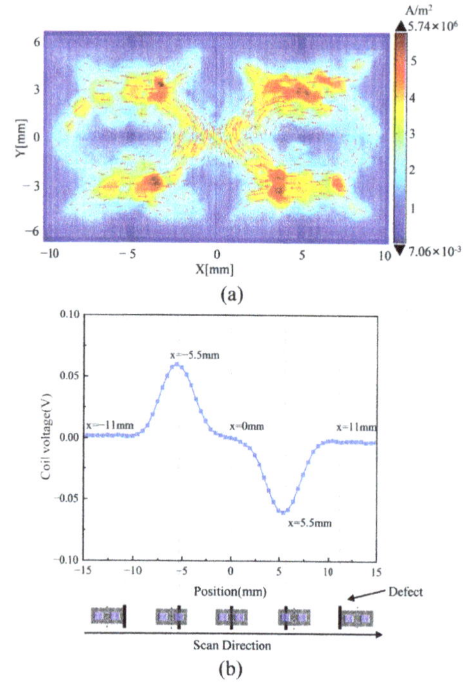

Figure 3. Simulation analysis of eddy current distribution and sensor response at 1 MHz: (**a**) Eddy current distribution on the test specimen. (**b**) Variation in sensor response when scanning for defects.

2.3. Probe Optimization

As shown in Section 2.2, the diamond-shaped eddy current distribution on the surface of the specimen is caused by the overlapping of the two excitation coils. Changing the angle of the two excitation coils will alter the eddy current distribution, which will impact the detected signals. Additionally, the detected signal may be impacted by the angle formed by the excitation and receiver coils, and optimization is performed in this section. As illustrated in Figure 4, the angle α is defined as the superposition angle between the upper and lower excitation coils, which governs the distribution of eddy currents on the specimen surface. The angle β is defined as the superposition angle between the excitation and receiver coils, which determines the sensitivity of the detection signals. Both α and β are 0° in the probe configuration depicted in Section 2.2.

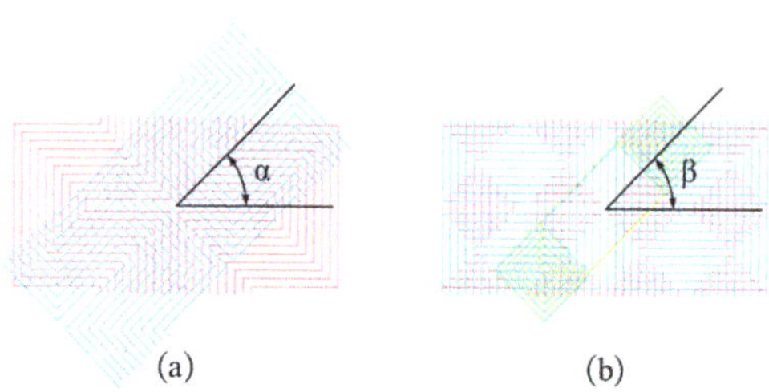

Figure 4. Schematic diagram illustrating the variation in coil angles: (**a**) Excitation coil overlap angle. (**b**) Receiving coil angle.

A simulation model similar to Figure 2 is used in this section. Firstly, the orientation angle α is optimized. Due to the structural symmetry, α is varied from 0° to 90° in 15° increments.

The corresponding results are shown in Figure 5. As shown in Figure 5a, when $\alpha = 0°$, the eddy current density reaches its peak, with a maximum value of 1.64×10^7 A/m^2; the induced eddy currents form a highly uniform and concentrated diamond-shaped pattern centered beneath the probe, indicating strong and effective excitation. As the angle α increases, both the intensity and the uniformity of the eddy current distribution gradually degrade. As shown in Figure 5b–d, compared with the 0° defect, all angled defects exhibit a reduced eddy current density and increased asymmetry in their distribution. Specifically, the maximum eddy current density decreases to 1.27×10^7A/m^2 at 30°, 9.41×10^6A/m^2 at 60°, and 1.18×10^7A/m^2 at 90°. The eddy current progressively deviates from the central axis, and the overall field strength weakens significantly. At larger angles, the eddy current becomes more dispersed, indicating a decline in detection sensitivity. Furthermore, as illustrated in Figure 6a, the induced voltage in the receiving coil is highest at $\alpha = 0°$, confirming that this orientation provides optimal excitation and detection performance.

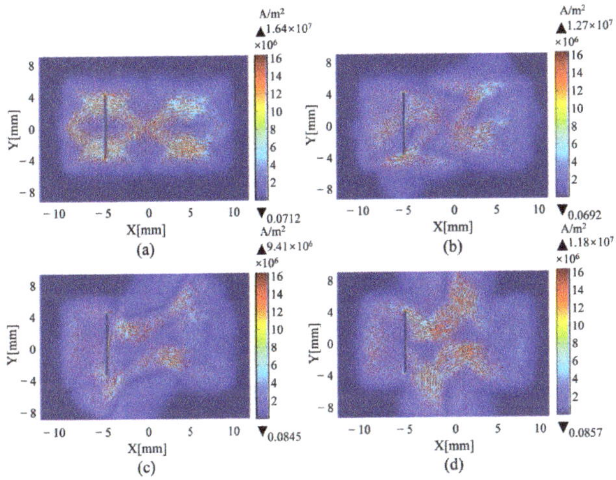

Figure 5. Eddy current density distribution with varying overlapping angles of upper and lower excitation coils: (**a**) 0°; (**b**) 30°; (**c**) 60°; (**d**) 90°.

Subsequently, the optimization of the receiver coil angle β is performed. β is varied from 0° to 90° in 15° increments. The results are shown in Figure 6b. The receiver coil voltage is maximized when β is set to 0°, which provides additional evidence that the best detection performance is achieved when both the excitation and receiver coil angles are set to 0°.

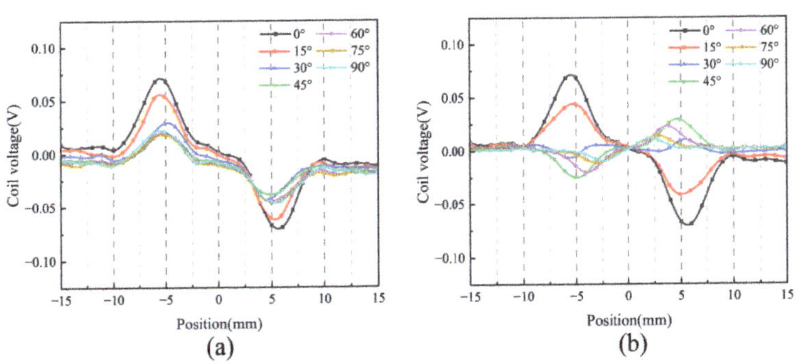

Figure 6. Relationship between coil voltage signal and coil angle variation: (**a**) Excitation coil overlap angle. (**b**) Receiving coil angle.

3. Simulation Analysis of FC with Different Angles

To evaluate the probe's capability in detecting defects with varying angles, simulations are conducted with defect angles set at 0°, 30°, 60°, and 90°, using a defect size 0.35 mm in width, 8 mm in length, and 2 mm in depth. The defect dimensions are consistent with those used in Section 2. Eddy current density is used for analysis, and it is defined as the difference between the eddy current distributions on the specimen surface with and without defects. Figure 7 presents the eddy current density with defect angles. Although the maximum eddy current density values are relatively consistent across all defect angles (1.09×10^7 to 1.18×10^7 A/m^2), the overall morphology of the eddy current distribution exhibits pronounced variations. The induced voltage signals, shown in Figure 8a, further confirm this trend. The highest response is observed at a defect angle of 90°, and the voltage amplitude gradually decreases as the angle decreases, highlighting the probe's sensitivity variation with respect to defect orientation.

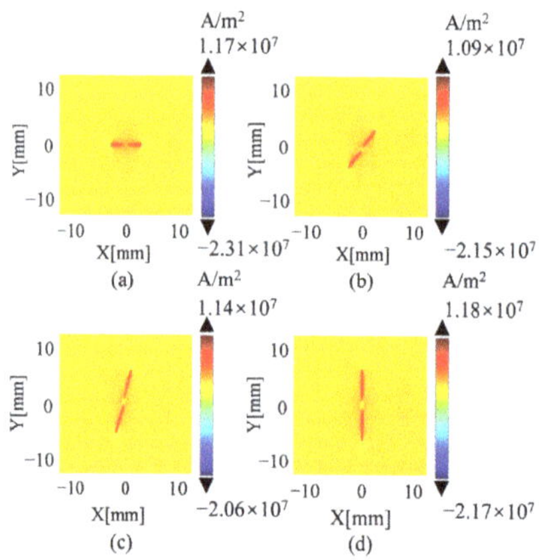

Figure 7. Eddy current density diagrams for the designed coil at different defect angles: (**a**) 0°; (**b**) 30°; (**c**) 60°; (**d**) 90°.

To further investigate the influence of defect angle on the detection signal, two time-domain features, ΔP and ΔW, are introduced [29]. ΔP denotes the peak-to-peak voltage, while ΔW represents the signal width, defined as the absolute difference between the *x*-coordinates corresponding to the peak-to-peak voltage, as illustrated by the red and blue markers in Figure 8a. Figure 8b presents the scatter plots and fitted curves of ΔP and ΔW as functions of the defect angle. As shown, both features exhibit clear dependencies on the defect angle, providing quantitative insight into how signal characteristics evolve with defect orientation. Specifically, ΔP increases progressively with defect angle, indicating greater signal variation due to changes in defect–probe interaction geometry. In contrast, ΔW decreases with increasing defect angle and gradually stabilizes, which can be attributed to a shorter interaction duration between the probe and the defect at larger angles.

To assess the effectiveness of the proposed flexible probe, its performance is compared with a commonly used rectangular coil under identical conditions. Figure 9 shows the rectangular coil structure, which has the same dimensions and wire diameter as the designed coil, and employs the same differential receiving coil to ensure consistent measurement conditions. Figure 10a presents the voltage signals of the designed coil at various defect angles, and Figure 10b displays those of the rectangular coil, including the scatter plot and fitted curve of ΔP and ΔW. The rectangular coil's ΔP varies non-monotonically with defect angle, likely due to uneven eddy current distribution, which complicates accurate angle discrimination where signal amplitudes overlap. In contrast, the designed coil exhibits a monotonic ΔP increase, offering better resolution. Both show a decreasing ΔW with increasing angle. Overall, the proposed probe outperforms the rectangular coil in defect detection across angles due to its superior stability and discrimination.

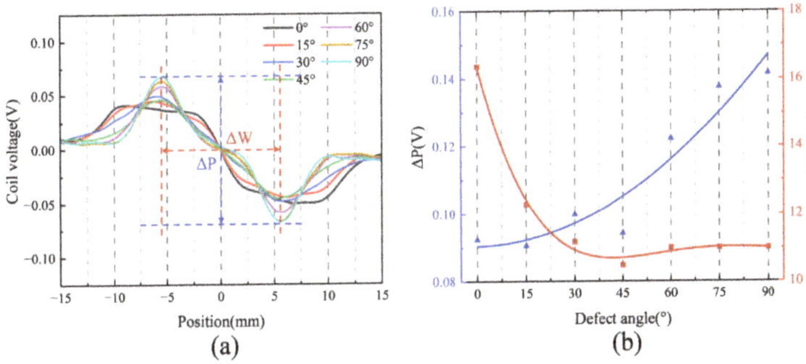

Figure 8. Coil voltage signals of the designed coil at different defect angles: (**a**) Comparison of test results. (**b**) Characteristic quantities.

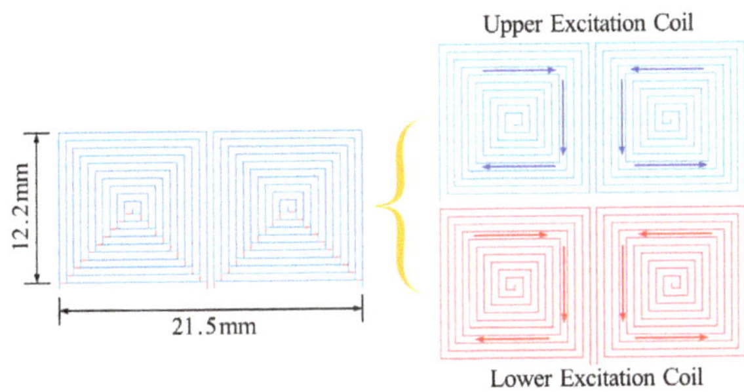

Figure 9. Rectangular coil structure design.

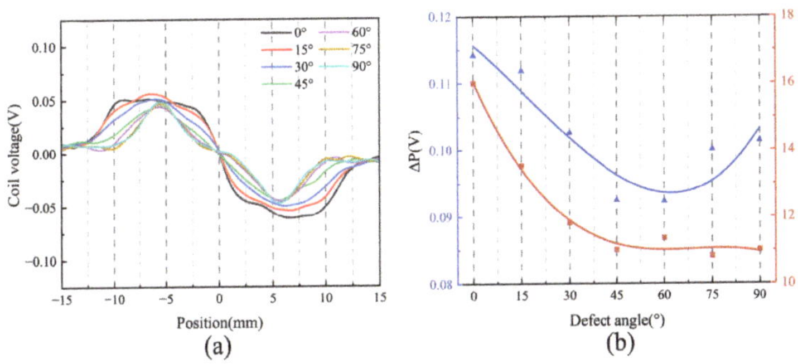

Figure 10. Coil voltage signals of the rectangular coil at different defect angles: (**a**) Comparison of test results. (**b**) Characteristic quantities.

4. Experimental Setup

The experimental setup is shown in Figure 11, where Figure 11a illustrates the experimental platform, and Figure 11b presents a physical representation. It comprises a signal generator, a power amplifier, a signal conditioning unit, a data collecting card, a scanning rack, a computer, an FPCB probe, and a flat panel specimen. The signal generator (DG4062, Rigol Technologies, Inc., Beijing, China) is used to generate two 1 MHz sine wave signals with an amplitude of 16 Vpp. One of the two generated signals is fed through the power amplifier to drive the excitation coil, providing a current of about 500 mA. The other signal, at the same frequency, is used as a reference for the signal conditioning unit.

This reference signal is subsequently processed through a phase shifter to produce two orthogonal components, enabling precise quadrature demodulation. The power amplifier (Model 7602M, Krohn-Hite Corporation, Brockton, MA, USA) is used to amplify one of the signals and deliver it to the excitation coil of the FPCB probe to generate the primary magnetic field. The signal conditioning unit consists of a preamplifier (OPA1611, G = 11), a phase shifter (OPA1652, G = 1), and a quadrature demodulation module (AD630 + NE5532, G = 2). It is used to generate orthogonal reference signals from the second output of the signal generator and to process the alternating signal induced in the receiving coil. The total system gain is approximately 22, enabling effective amplification of weak sensor signals while maintaining a high signal-to-noise ratio. A fourth-order Butterworth low-pass filter is applied to suppress high-frequency components, reducing noise and improving overall signal quality. The data acquisition card (NI-6351, National Instruments Corporation, Austin, TX, USA) is used to synchronously collect the processed signals and transfer them to the computer via DAQExpress 5.0 software for storage and analysis. The scanning rack is used to hold and move the FPCB probe across the surface of the specimen during experiments, enabling automated scanning. As shown in Figure 11b, the FPCB probe consists of a double-layer excitation coil and a differential receiving coil. The detailed parameters of the probe are given in Table 2. At an excitation frequency of 1 MHz, the average resistance and inductance of the sensing units are 14.22 Ω and 4.28 μH, with fluctuations of about 0.5% and 0.7%, respectively.

Table 2. Geometric parameters of the FPCB probe.

Coil	Wire Diameter (mm)	Wire Spacing (mm)	Wire Turns	Unit Width (mm)	Lateral Width (mm)
Excitation coil	0.1	0.1	26	\	47.85
Receiving coil	0.1	0.1	13	5.6	23.6

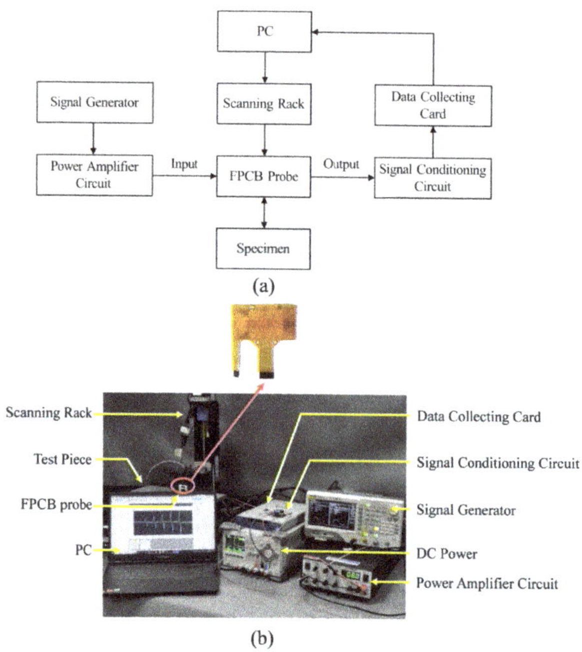

Figure 11. Experimental platform: (**a**) Schematic diagram of the experimental platform. (**b**) Physical drawing of the experimental platform.

In addition, the aluminum samples with defects are presented in Figure 12. Flat aluminum specimens with dimensions of 600 × 200 × 8 mm are used in the experiments. A series of artificial defects are introduced using Electrical Discharge Machining (EDM). These defects are uniformly spaced and increase progressively in size from left to right. To prevent interference between adjacent defects during testing, the probe scans the specimen along three predefined paths, denoted as paths a–c. Each path contains five defects, resulting in a total of fifteen defects. The detailed dimensions of the defects are listed in Table 3.

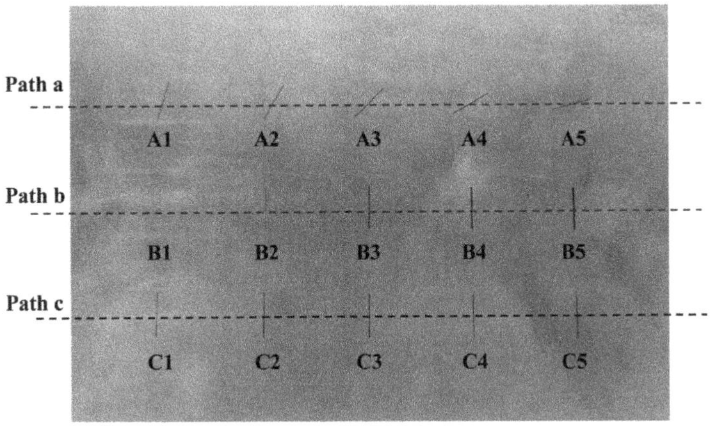

Figure 12. Aluminum sample with artificial defects.

Table 3. Geometric parameters of defects.

Defect Number	Length (mm)	Width (mm)	Depth (mm)	Angle (°)
A1				75
A2				60
A3	20	0.35	2	45
A4				30
A5				15
B1		0.10		
B2		0.20		
B3	20	0.35	2	90
B4		0.50		
B5		0.70		
C1			0.50	
C2			0.75	
C3	20	0.35	1.00	90
C4			1.50	
C5			2.00	

5. Discussion

5.1. Experimental Validation of ΔP and ΔW for Defect Geometry Characterization

To validate the simulation results presented in Section 3, experiments are conducted along the predefined scan path a using the experimental platform described in Section 4 to evaluate the probe's response to defects with varying angles. Each defect is measured ten times under identical conditions, including a lift-off of 0.5 mm and a scanning speed of 1 mm/s. The sampling rate of the data acquisition card is 2000 samples/s. However, the raw signals from the data acquisition card are not used directly due to random errors caused by data acquisition and baseline drift induced by temperature. To address this, median filtering and detrending methods are applied, with the latter involving subtraction of an optimal least-squares fitting line to eliminate baseline drift. The repeated measurements are then used to obtain the mean and standard deviation of the characteristic parameters ΔP and ΔW, which serve to evaluate repeatability and experimental uncertainty.

Figure 13 presents the experimental results for five defects with different angles. As shown in Figure 13a, the probe successfully detects all five defects, producing distinct signal responses characterized by variations in both amplitude and width, which are consistent with the simulation results. The responses also exhibit a high signal-to-noise ratio. Specifically, Figure 13b demonstrates that ΔP increases monotonically with increasing defect angle, aligning well with the simulation data presented in Figure 8. Moreover, ΔW decreases with increasing defect orientation, in general agreement with simulation results. Simulations indicate that ΔW tends to plateau at higher orientations, whereas experimental measurements show a continued decrease at the largest orientations, albeit with a diminished rate. Despite this minor deviation, the findings confirm that ΔW remains a robust feature sensitive to defect orientation.

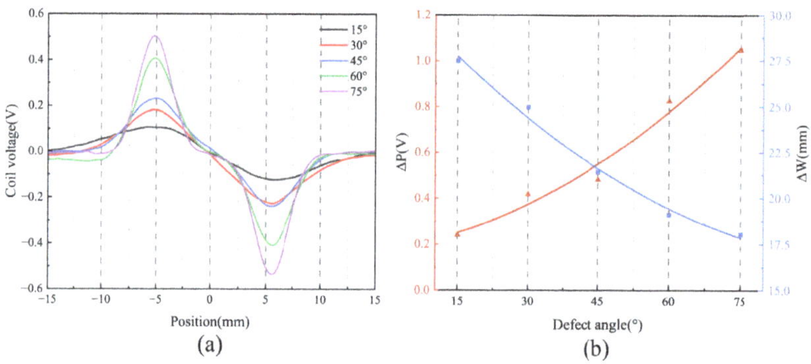

Figure 13. Defect detection results for defects of different angles: (**a**) Comparison of test results. (**b**) Characteristic quantities.

To further assess the probe's capability in detecting defects of varying sizes, experiments are conducted along the predefined scan paths b and c.

Figure 14 illustrates the probe's response to defects of varying widths. As shown in Figure 14a, the black curves represent the response corresponding to the 0.1 mm defect, indicating that the probe reliably detects narrow cracks as small as 0.1 mm and demonstrates high spatial resolution. In Figure 14b, the parameter ΔP increases monotonically with defect width; this is because wider defects cause greater disruption to eddy currents, resulting in stronger signals. In contrast, ΔW remains approximately 20 mm with minor fluctuations, indicating low sensitivity to variations in defect width. This behavior contrasts with the angle experiments, where ΔW varied significantly. This is because variations in defect angle substantially change the orientation of the eddy current perturbation relative to the probe's sensing field, thereby modifying the temporal evolution of the induced signal and producing pronounced changes in ΔW. In contrast, defect width mainly alters the lateral spread of the perturbation within the material, which exerts only a limited influence on the temporal width of the signal and thus results in minimal variation in ΔW.

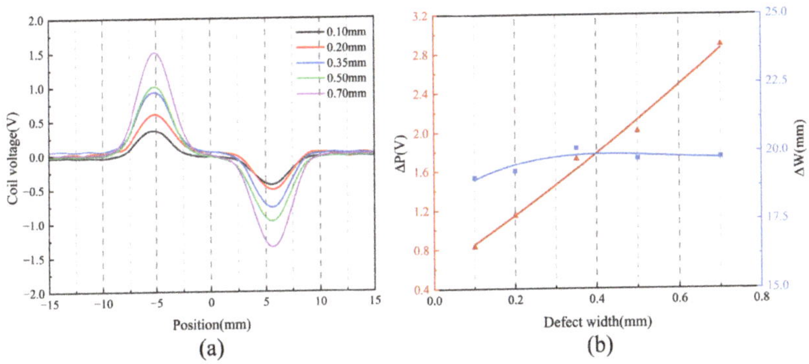

Figure 14. Defect detection results for defects of different widths: (**a**) Comparison of test results. (**b**) Characteristic quantities.

Figure 15 presents the probe's response to defects of varying depths. As shown in Figure 15a, the black curves represent the response corresponding to the 0.5 mm defect, demonstrating the probe's ability to detect subsurface cracks as shallow as 0.5 mm with high sensitivity to vertical variations. In Figure 15b, as defect depth increases, the parameter ΔP rises proportionally, reflecting a greater disruption of the eddy current field. Conversely, ΔW remains approximately 20 mm with minor fluctuations, indicating low sensitivity to variations in defect depth. ΔW response to depth variations likely arises because depth primarily affects the intensity of the eddy current disturbance, with negligible impact on its spatial or temporal profile, resulting in a stable temporal width measurement.

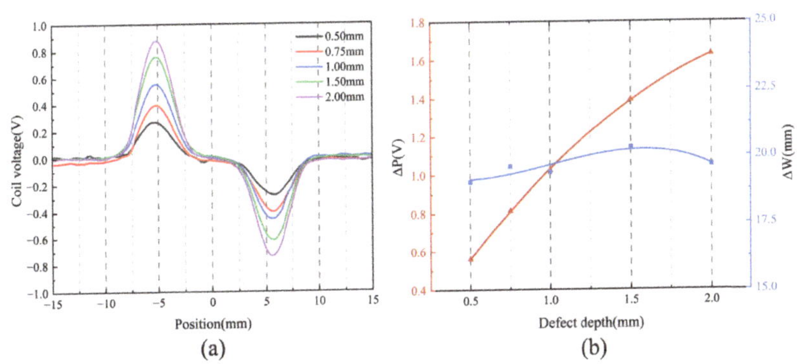

Figure 15. Defect detection results for cracks of different depths: (**a**) comparison of test results and (**b**) characteristic quantities.

The results demonstrate that ΔP is sensitive to variations in all defect parameters, including angle, width, and depth, exhibiting noticeable changes across these dimensions. Consequently, ΔP alone cannot definitively identify which specific defect dimension has changed. Similarly, ΔW primarily responds to changes in defect angle and exhibits limited sensitivity to width and depth. Notably, the variation ranges of ΔW corresponding to different defect parameters overlap significantly: 17.6 mm to 27.5 mm for angle, and approximately 20 mm for both width and depth. Due to this overlap, ΔW also fails to accurately distinguish the specific defect parameter, which causes difficulties in quantitative analysis. Therefore, it is necessary to classify defects prior to quantitative analysis.

5.2. Transformer-Based Autoencoder for Defect Feature Analysis

A Transformer-based autoencoder framework is introduced to classify defect types, thereby facilitating subsequent quantitative analysis. The Transformer is a neural network architecture based on self-attention, capable of capturing long-range dependencies and subtle temporal variations in sequential data, making it well suited for analyzing complex eddy current signals. Its parallel processing capability and ability to model non-local interactions contribute to its effectiveness in defect classification tasks.

Time-domain eddy current signals are collected from three defect types (angle, width, depth), with five representative signals per type, yielding 15 original signals. The dataset is expanded tenfold via data augmentation, including Gaussian noise injection, amplitude scaling, and time shifting. All signals are resampled to a uniform length and standardized, and auxiliary features—defect length, width, depth, angle, and two time-domain characteristics (ΔP, ΔW)—are also standardized. The signals are projected into a 32-dimensional embedding space and combined with auxiliary features to form the model input.

This combined input is processed through a single-layer Transformer encoder with four self-attention heads and a feedforward dimension of 128, ultimately producing a 128-dimensional latent representation for defect classification. The model is trained with five-fold cross-validation over 150 epochs using the Adam optimizer (learning rate 1×10^{-4}, weight decay 1×10^{-4}), with a composite loss function ($0.7 \times$ MSE + $0.3 \times$ cross-entropy). Gradient accumulation over four steps is used to reduce memory consumption. The latent vector is then mapped through fully connected layers to generate final defect class probabilities, and the model with the lowest validation loss across all folds is selected for evaluation.

To establish a performance baseline, a CNN-based model is implemented with comparable model complexity and identical training settings as the Transformer. The CNN consists of three convolutional blocks with channels of 32, 64, and 128, using kernel sizes of 7, 5, and 3, each followed by ReLU activation and max-pooling. After adaptive average pooling to size 32, the features are flattened and concatenated with separately processed auxiliary features (6-dimensional input mapped to 64 dimensions), then projected to a 128-dimensional latent space matching the Transformer's feature dimensionality. Both models are evaluated using five-fold cross-validation with identical data splits. The training and validation loss curves are shown in Figure 16. Both models show a steady decline in loss that levels off toward the end of training, indicating smooth convergence and stable training. The performance metrics for both models are listed in Table 4. The results indicate that the Transformer achieves slightly higher accuracy and precision compared with the CNN, demonstrating a consistent performance improvement across all metrics. Although the Transformer requires a somewhat longer training time due to its computational complexity, this additional cost is justified by the improved classification accuracy and robustness under varying inspection conditions. Furthermore, the Transformer framework can readily integrate auxiliary features such as frequency-domain characteristics or imaging data, which will be valuable for more complex defect characterization in future work.

Table 4. Performance comparison of the Transformer and CNN models.

Model	Epoch	Accuracy (%)	Precision (%)	Training Time (s)
Transformer	150	98.0	98.2	55.5
CNN	150	97.3	97.7	37.6

To visually assess how well the Transformer autoencoder captures differences between defect types, the 128-dimensional latent features are first reduced to 50 dimensions using Principal Component Analysis (PCA) and subsequently clustered with KMeans, with the optimal number of clusters determined by the silhouette score. The features are then projected into two dimensions using t-distributed Stochastic Neighbor Embedding (t-SNE), resulting in well-separated clusters that correspond closely to the different defect types. Figure 17 shows the clustering results of different defect types. The visualized clusters exhibit clear separation and compactness, corresponding well with defect categories such as angle, width, and depth. This indicates that the proposed feature extraction and dimensionality reduction approach effectively captures the distinguishing characteristics of various defect types.

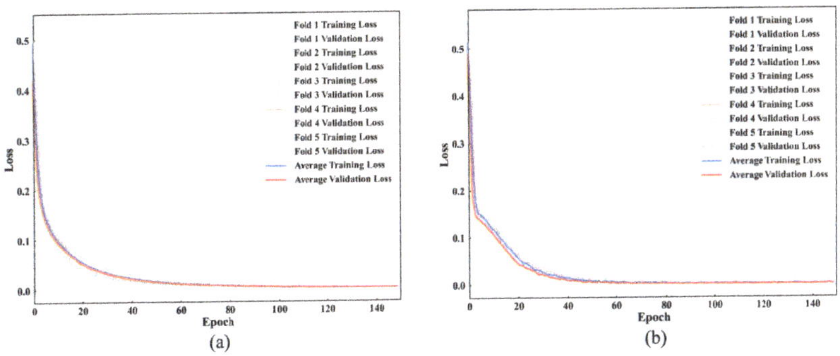

Figure 16. Training and validation loss curves: (**a**) Transformer and (**b**) CNN.

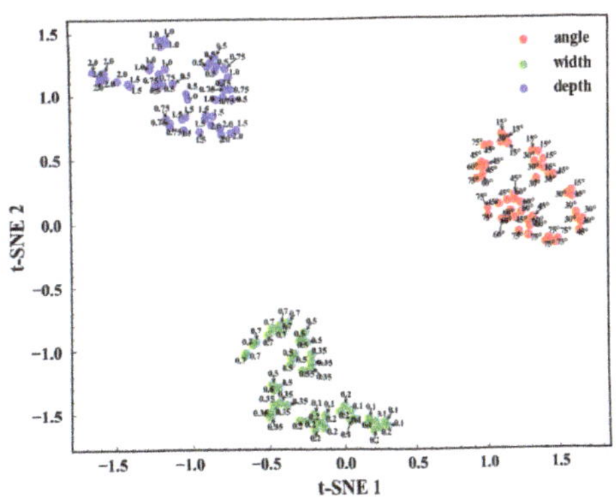

Figure 17. t-SNE visualization of the clustered latent features for different defect types.

For practical defect detection, defect type identification should be performed as the preliminary step prior to quantitative analysis, since determining the defect category in advance allows the model to apply defect-specific regression strategies, thereby improving both the accuracy and efficiency of subsequent dimensional characterization.

6. Conclusions

The proposed FPCB probe demonstrates enhanced eddy current inspection performance through its double-layer planar excitation and differential receiving coil design. The excitation coil employs a reverse-wound configuration to enhance both the directionality and focusing of the magnetic field, while the differential receiving coil improves sensitivity and suppresses common-mode noise. Optimization of both the excitation and receiving coil angles maximizes eddy current density and detection signal strength. Moreover, finite element simulations and experimental validations show that the probe achieves high spatial resolution and robust sensitivity across a variety of surface cracks, while analysis of the time-domain features, ΔP and ΔW, shows that although they are sensitive to various defect parameters, overlapping responses necessitate a Transformer-based classification strategy for accurate quantitative evaluation. The proposed method demonstrates reliable performance and clear interpretability for defect evaluation in aluminum components.

Author Contributions: Conceptualization, Q.Z. and H.W.; methodology, Q.Z. and J.Z.; formal analysis, J.Z. and Y.W.; investigation, S.W.; resources, H.W.; writing—original draft, J.Z. and Y.W.; writing—review and editing, Q.Z., L.L. and H.W.; supervision, Q.Z. and H.W. All authors have read and agreed to the published version of the manuscript.

Funding: This work was supported by the National Natural Science Foundation of China under Grant [61903193]; the Science and Technology Program of the State Administration for Market Regulation under Grant [2024MK038]; the Cultivation Fund for the Joint Laboratory of New Nondestructive Testing Technologies between Nanjing University of Aeronautics and Astronautics and the University of Lisbon; the Nondestructive Detection and Monitoring Technology for High Speed Transportation Facilities, Key Laboratory of Ministry of Industry and Information Technology; the Jiangsu Provincial Key Laboratory of Culture and Tourism for Nondestructive Testing and Safety Traceability of Cultural Relics; and the Fundamental Research Funds for the Central Universities under Grants [NJ2024015] and [NJ2024045].

Institutional Review Board Statement: Not applicable.

Informed Consent Statement: Not applicable.

Data Availability Statement: Dataset available on request from the authors.

Conflicts of Interest: Author Yanchang Wang was employed by the company China State Shipbuilding Corporation Limited. The remaining authors declare that the research was conducted in the absence of any commercial or financial relationships that could be construed as a potential conflict of interest.

References

1. Li, S.S.; Yue, X.; Li, Q.Y.; Peng, H.L.; Dong, B.X.; Liu, T.S.; Yang, H.Y.; Fan, J.; Shu, S.L.; Qiu, F.; et al. Development and applications of aluminum alloys for aerospace industry. *J. Mater. Res. Technol.* **2023**, *27*, 944–983. [CrossRef]
2. Czerwinski, F. Aluminum alloys for electrical engineering: A review. *J. Mater. Sci.* **2024**, *59*, 14847–14892. [CrossRef]
3. Huang, G.; Li, Z.H.; Sun, L.M.; Li, X.W.; Wen, K.; Yan, L.Z.; Xiong, B.Q.; Zhang, Y.A. Fatigue crack growth behavior of 2624-T39 aluminum alloy with different grain sizes. *Rare Met.* **2021**, *40*, 2523–2529. [CrossRef]
4. Zhang, L.; Zhong, H.; Li, S.; Zhao, H.; Chen, J.; Qi, L. Microstructure, mechanical properties and fatigue crack growth behavior of friction stir welded joint of 6061-T6 aluminum alloy. *Int. J. Fatigue* **2020**, *135*, 105556. [CrossRef]
5. Mughrabi, H. Microstructural mechanisms of cyclic deformation, fatigue crack initiation and early crack growth. *Philos. Trans. R. Soc. A Math. Phys. Eng. Sci.* **2015**, *373*, 20140132. [CrossRef] [PubMed]
6. García-Martín, J.; Gómez-Gil, J.; Vázquez-Sánchez, E. Non-destructive techniques based on eddy current testing. *Sensors* **2011**, *11*, 2525–2565. [CrossRef]
7. Xie, R.; Chen, D.; Pan, M.; Tian, W.; Wu, X.; Zhou, W.; Tang, Y. Fatigue crack length sizing using a novel flexible eddy current sensor array. *Sensors* **2015**, *15*, 32138–32151. [CrossRef]
8. Gupta, M.; Khan, M.A.; Butola, R.; Singari, R.M. Advances in applications of Non-Destructive Testing (NDT): A review. *Adv. Mater. Process. Technol.* **2022**, *8*, 2286–2307. [CrossRef]
9. Song, M.; Qiu, Y.; Xu, B.; Jiang, P.; Geng, S.; Wang, Y.; Zhao, J.; Hu, Y. Fatigue failure and grain refinement strengthening mechanism of aluminum alloy weld. *Int. J. Fatigue* **2025**, *197*, 108950. [CrossRef]
10. Borrego, L.; Jesus, J.; Branco, R.; Ferreira, J.; Antunes, F.; Neto, D.; Sérgio, E. Mechanisms of fatigue crack growth in 7050-T6 aluminium alloy. *Int. J. Fatigue* **2025**, *194*, 108830. [CrossRef]
11. Machado, M.A. Eddy currents probe design for NDT applications: A review. *Sensors* **2024**, *24*, 5819. [CrossRef] [PubMed]
12. Mussatayev, M.; Yi, Q.; Fitzgerald, M.; Maes, V.K.; Wilcox, P.; Hughes, R. Directional eddy current probe configuration for in-line detection of out-of-plane wrinkles. *Compos. Part B Eng.* **2024**, *268*, 111048. [CrossRef]
13. Fan, X.; Chen, T.; Du, J.; He, Y.; Cui, R.; Ma, B.; Song, Y. Methods for improving sensitivity of crack quantitative monitoring of flexible eddy current array sensor. *Smart Mater. Struct.* **2020**, *29*, 085033. [CrossRef]

14. Wu, D.; Cheng, F.; Yang, F.; Huang, C. Non-destructive testing for carbon-fiber-reinforced plastic (CFRP) using a novel eddy current probe. *Compos. Part B Eng.* **2019**, *177*, 107460. [CrossRef]
15. Wang, Y.; Ye, C.; Wang, M. Synthetic magnetic field imaging with triangle excitation coil for inspection of any orientation defect. *IEEE Trans. Instrum. Meas.* **2019**, *69*, 533–541. [CrossRef]
16. Ge, J.; Yang, C.; Yu, F.; Yusa, N. Transformation of the rotating eddy current testing signal at the desired eddy current orientation. *NDT E Int.* **2022**, *125*, 102551. [CrossRef]
17. Ge, J.; Hu, B.; Yang, C.; Yu, F.; Yusa, N. Surface profile reconstruction of complex cracks using the signals of rotating eddy current testing through the eddy current imaging method. *IEEE Trans. Ind. Electron.* **2022**, *70*, 9632–9641. [CrossRef]
18. Xu, Z.; Wang, X.; Deng, Y. Rotating focused field eddy-current sensing for arbitrary orientation defects detection in carbon steel. *Sensors* **2020**, *20*, 2345. [CrossRef]
19. Trung, L.Q.; Kasai, N.; Sekino, K.; Miyazaki, S. An eddy current convergence probe with copper core and single detection coil to detect flaws on aluminum plates. *NDT E Int.* **2022**, *132*, 102707. [CrossRef]
20. Trung, L.Q.; Kasai, N.; Sekino, K.; Miyazaki, S. Eddy current convergence probes with self-differential and self-nulling characteristics for detecting cracks in conductive materials. *Sens. Actuators A Phys.* **2023**, *349*, 114084. [CrossRef]
21. Liu, Z.; Li, Y.; Ma, W.; Tang, Z.; He, C. A novel magnetic flux-convergence semi-ellipsoidal eddy current probe for defect detection. *IEEE Sens. J.* **2025**, *25*, 19001–19013. [CrossRef]
22. Wang, G.; Zhong, Y.; Xiao, Q.; Li, Q. Eddy current sensor with multiple spatial degrees of freedom for concentrating magnetic fields. *IEEE Trans. Instrum. Meas.* **2023**, *72*, 6007711. [CrossRef]
23. Li, X.; Tian, G.; Li, K.; Wang, H.; Zhang, Q. Differential ECT probe design and investigation for detection of rolling contact fatigue cracks with different orientations. *IEEE Sens. J.* **2022**, *22*, 11615–11625. [CrossRef]
24. She, S.; He, Y.; Chen, Y.; Chady, T. Flexible floral eddy current probe for detecting flaws in metal plate. *IEEE Sens. J.* **2020**, *20*, 10521–10529. [CrossRef]
25. Chen, G.; Li, Z.; Cao, Z.; Gao, W.; Jin, W. A differential excitation eddy current probe based on novel annular fractal curve for defects inspection. *IEEE Sens. J.* **2022**, *22*, 15903–15915. [CrossRef]
26. Chen, G.; Wei, J.; Zhang, S.; Gao, W.; Jin, W.; Fan, L. Differential Koch planar eddy current probe sensitivity boosting by iron particulate sheet. *IEEE Trans. Instrum. Meas.* **2023**, *72*, 6010609. [CrossRef]
27. Trung, L.Q.; Kasai, N.; Le, M.; Sekino, K. Predicting actual crack size through crack signal obtained by advanced Flexible Eddy Current Sensor using ResNet integrated with CBAM and Huber loss function. *NDT E Int.* **2025**, *149*, 103249. [CrossRef]
28. Ma, Q.; Gao, B.; Tian, G.; Yang, C.; Xie, L.; Chen, K. High sensitivity flexible double square winding eddy current array for surface micro-defects inspection. *Sens. Actuators A Phys.* **2020**, *309*, 111844. [CrossRef]
29. Arismendi, N.O.R.; Pacheco, E.R.; López, O.P.; Espina-Hernández, J.H.; Benitez, J.A.P. Classification of artificial near-side cracks in aluminium plates using a GMR-based eddy current probe. In Proceedings of the 2018 International Conference on Electronics, Communications and Computers (CONIELECOMP), Cholula, Mexico, 21–23 February 2018; pp. 31–36. [CrossRef]

sensors

Article

Comparison Research on Characterization and Evaluation Approaches for Paint Coated Corrosion Using Eddy Current Pulsed Thermography

Chao Xue [1], Yinqiang Zhang [1,*], Song Ding [1,2,*], Cheng Song [3] and Yiqing Wang [1]

1 School of Electrical Engineering and Control Science, Nanjing Tech University, Nanjing 211816, China; 202061206103@njtech.edu.cn (C.X.); wangyiqing@njtech.edu.cn (Y.W.)
2 Key Laboratory of Nondestructive Testing, Fujian Polytechnic Normal University, Fujian Province University, Fuqing 350300, China
3 Siemens Limited China, Beijing 100102, China; 13911899416@139.com
* Correspondence: zhang_blue@njtech.edu.cn (Y.Z.); dingsong@njtech.edu.cn (S.D.)

Abstract: Paint coated corrosion detection and evaluation is a big challenge for steel performance and structure health. Eddy current pulsed thermography (ECPT) technique is investigated because it can reflect the corrosion physical properties through paint coating by the infrared signal. This paper proposes skewness method, which presents the feature of temperature curve's shape automatically, and compares it with principal component analysis (PCA), phase analysis, and kurtosis feature extraction methods for paint coated corrosion characterization and evaluation. The averaged skewness shows the best sensitivity for 0–6 months corrosion. The normalized second principal component (PC) presents good sensitivity and the best measurement scale for corroded time. Furthermore, the temperature curve analysis proves that the electrical conductivity dominates the induced heating and heat distribution. The corrosion height is utilized to explain why ECPT technique is valid within 10 months corroded time. ECPT technique is proved as a smart sensor system for paint coated corrosion detection and characterization.

Keywords: corrosion evaluation methods; paint coated corrosion; eddy current pulsed thermography; smart sensor system

1. Introduction

Steel is a fundamental material for our life and industry. However, corrosion, the wastage and destruction caused by chemical or electrochemical reactions with surroundings, weakens steel's performance and damages the structure's health, even human lives [1,2]. For corrosion measurement, many physical, chemical, and electrochemical methods were employed, such as mass loss measurement, scanning Kelvin probe (SKP), ultrasonic testing (UT), scanning acoustic microscopy (SAM), electrochemical impedance spectroscopy (EIS), electrochemical noise (EN), and so on [3–6]. With the wide application of the paint coating technique, which is provided to protect steel or structure surfaces, it is more difficult to detect the local corrosion under the paint coating. In order to non-destructively detect local corrosion under paint coatings, there are some non-destructive testing techniques, such as the Barkhausen noise method, eddy current testing method, magnetic field measurement method, etc. However, these technologies are susceptible to factors such as material magnetoelasticity, texture, and stress distribution, making signal analysis more complex, and eddy current testing can only detect defects on the surface or near the surface of the material. Eddy current pulse thermal imaging technology can utilize multidimensional feature fusion algorithms to improve defect recognition rate and quantitative evaluation ability [7].

Recently, the ECPT technique is engrossing as well as the corresponding signal processing methods [8–10]. For this technique, an AC supply is employed to drive an excitation magnetic field, which generates induced eddy current and Joule heat on the surface of the conductive material. After a short time, the power supply is switched off and the specimen's temperature falls to the initial value. The two stages, including Joule heat generation and temperature decline are recorded by an infrared camera. As the infrared emission could reflect the physical properties and their variation of the specimen, the ECPT technique was investigated for nondestructive testing and characterization for defects, such as cracks and corrosion blister. This novel technique has the advantages of non-contact and mobile, big lift-off distance, high sensitivity and resolution, which contributes to on-site detection and characterization. In particular, the excited magnetic field can induce an eddy current on the steel through a nonconductive paint layer, which makes it possible to detect defects under paint coating.

For processing the infrared images, transient analysis methods and statistic algorithms are investigated. M. Pan et al. [11] employed PCA and independent component analysis (ICA) [12] methods to characterize the delamination defect of carbon fiber reinforced plastic (CFRP). The PC and independent components were extracted as the features for thermal image reconstruction and damage characterization. Y. He et al. [13] compared PCA, ICA and fast Fourier transform (FFT) in the performance of blister detection. In conclusion, the third PC, and the first and fourth independent components could be used to detect corrosion blisters. The FFT algorithm could eliminate the non-uniform heating and detect the blister easily. R. Yang et al. [14] analyzed the process of joule heat generation, conduction and radiation on the corroded and uncorroded regions. Accordingly, a heat transport model and phase analysis approach were proposed. The phase of temperature variation for each pixel was calculated as the feature by the FFT method. The proposed approach could characterize the corrosion region and nondestructively evaluate different corrosion states from 1 month to 6 months.

Compared with the transient methods, phase analysis extracts the features in the frequency domain based on a single pixel, which eliminates the influence of non-uniform heating and emissivity difference of different locations on the sample surface. However, the phase deference is theoretically limited between $-\pi \sim +\pi$, and the frequency selection for analysis is manual. In order to provide an approach for practice, which not only characterizes the corrosion with the reconstructed image but also quantitatively measures the corrosion stage. This paper aims to investigate and compare the feature extraction methods and proposes an approach based on the extracted features of pixels in the marked region for quantitative corrosion stage characterization and evaluation.

In the rest of this paper, the principle of the ECPT system and different four feature extraction algorithms are introduced in Section 2. The sample preparation and experimental setup are illustrated in Section 3. Next, in Section 4, the feature extraction and reconstruction are presented, and the corrosion characterization performance of the proposed four algorithms is compared. Based on the analysis and discussion, the conclusion and feature work is proposed in the last section.

2. Methodologies

ECPT is a multi-physics coupled nondestructive testing and evaluation technique, including induction heating, thermal conduction, infrared image acquisition, and data processing [8–10]. For the ECPT system, as shown in Figure 1, an alternating current drives the excitation coil to induce a high-frequency eddy current in the conductive samples. Then, Joule heat is generated in a short time by the inductive eddy current and resistance of the material. Then, the infrared image sequence is acquired by an infrared camera to reflect the heat generation, conduction, and diffusion processes [15]. On the other hand, the physical properties of materials, such as electrical conductivity, thermal conductivity, permeability and emissivity, are closely related to the microstructure of the material. Corrosion gradually changes the composition and microstructure of the material, which causes complex changes in electrical conductivity, and thermal conductivity [16,17], which is governed by Equations (1) and (2).

$$Q_1 = \frac{1}{\sigma_1}|J_s|^2,\ Q_2 = \frac{1}{\sigma_2}|J_s|^2 \quad (1)$$

$$\rho_1 C_{p1}\frac{\partial T}{\partial t} - \nabla(k_1 \nabla T) = Q_1,\ \rho_2 C_{p2}\frac{\partial T}{\partial t} - \nabla(k_2 \nabla T) = Q_2 \quad (2)$$

where Q_1 is the induced Joule heat of the corrosion layer, Q_2 is the induced Joule heat of the uncorroded layer, σ_1 is the electrical conductivity of the corrosive layer, σ_2 is the electrical conductivity of the uncorroded layer, J_s means eddy current density. ρ_1 is the material density of the corrosive layer, ρ_2 is the material density of the uncorroded layer, C_{p1} is the material heat capacity of the corrosive layer, C_{p2} is the material heat capacity of the uncorroded layer, k_1 is the thermal conductivity of the corrosive layer, k_2 is the thermal conductivity of the uncorroded layer. The degree of corrosion, as well as the properties of the products, can cause changes in conductivity and thermal conductivity. Therefore, the temperature variation indicates the state of the surface.

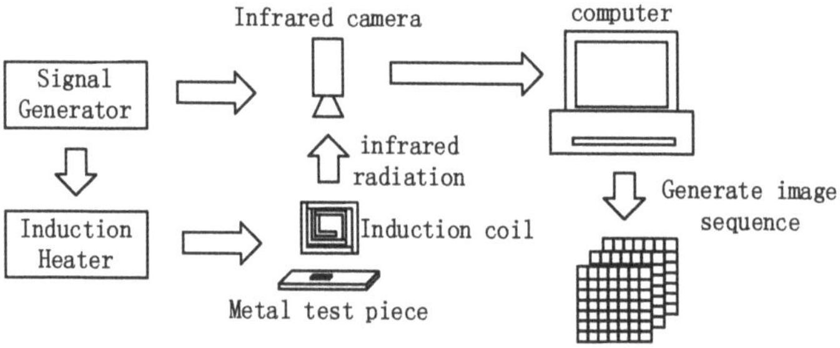

Figure 1. The diagram of ECPT system.

Considering the skin effect, described as Formula (3), the higher excitation frequency means the shallower depth of eddy current distribution, therefore, as well as the Joule heat distribution. Thus, the initial thermal layer can be estimated by δ [18]

$$\delta = \frac{1}{\sqrt{f\sigma\pi\mu}} \quad (3)$$

where f is the frequency of excitation, σ and μ are the electro-conductivity and permeability of the conductive material, respectively. Considering the excitation frequency and induction time, the initial heat flux on the sample surface is limited. Detailed information on these two parameters will be displayed in the experimental device.

Furthermore, due to the thermal image sequence including the noise of environmental temperature, sensor array, and sample geometry, feature extraction algorithms should be utilized to eliminate the test noise and enhance the accuracy.

2.1. Fast Fourier Transform

According to the model proposed by R. Yang [14], thermal waves are generated during heat conduction between corroded and free areas, and any waveform can be approximated by the sum of pure harmonics of oscillation at different frequencies. Due to the difference in the phase of the thermal pulse waveform between the corroded area and the uncorroded area, the Fourier transform algorithm can be used to extract the feature.

Fourier transform is a mathematical tool between the time domain and frequency domain. It is a commonly used method of analyzing signals, which essentially aggregates information from the time domain into the frequency domain. Any waveform, whether periodic or aperiodic, can be approximated by the sum of pure harmonics oscillating at different frequencies. In pulse eddy current thermal imaging, a pulse signal is applied to the coil, which excites the surface of the tested object to generate eddy currents. For an ideal pulse signal with zero duration, it can be decomposed into all frequency components of 0–∞, while the frequency components of the actual applied pulse signal are limited. The Joule heat generated by eddy current will also conduct thermal diffusion with these limited frequency components. The temperature distribution formed by diffusion is called a heat wave. According to the heat wave theory, the high-frequency heat wave propagates closer and faster, while the low-frequency heat wave propagates farther and slower. Therefore, low-frequency excitation can reveal deeper structural information within the tested object. However, the amplitude and phase of thermal waves are not only affected by pulse excitation within the tested object but also by changes in the thermal diffusion coefficient caused by defects. Therefore, defect information can be extracted from the amplitude-frequency and phase-frequency plots [19].

The infrared radiation signal collected by an infrared camera is a discrete signal. Therefore, a discrete Fourier transform is needed to decompose the signal into different frequency components. In this paper, the temperature response of each pixel is calculated according to the following one-dimensional discrete Fourier transform formula [20]:

$$F_n = \Delta t \sum_{k=0}^{N-1} T(k\Delta t) e^{-\frac{i2\pi nk}{N}} = Re_n + Im_n \quad (4)$$

where Δt is the sampling interval, n designates the frequency increment, N is the number of samples, Re_n and Im_n are, respectively, the real and imaginary components of F_n. Then, the amplitude and phase can be calculated by the following equation [20]:

$$A_n = \sqrt{Re_n^2 + Im_n^2} \quad (5)$$

$$\varphi_n = tan^{-1}\left|\frac{Im_n}{Re_n}\right| \quad (6)$$

Then, the amplitude and phase of all pixels at certain frequencies are extracted to reconstruct the thermal images.

2.2. Principal Component Analysis

Considering the data acquisition of the ECPT system, the infrared camera transforms the infrared emission into a thermal image sequence and transmits it to a computer for further analysis. For a single thermal image, the value for each pixel signifies the temperature of an independent region. Because any object that is above absolute zero will release infrared radiation, the measured temperature of each pixel should be considered as a random signal that includes the noise of the sensor unit and environmental temperature variation. Therefore, the thermal image sequence can be seen as a 3D random data set.

The PCA method was commonly used for reducing data dimension and noise [10]. Since the fundamental process of PCA is to calculate the orthorhombic directions in which the random data have obvious variances [20], as shown in Figure 2, this method is considered to have the potential to indicate the temperature gradient caused by defects.

Due to the corrosion changes in the electrical conductivity, thermal conductivity, and permeability, the temperature distribution around the defect is uneven. PCA extracts the image region information with large temperature gradients, which cause obvious deviation and contain the features of defects in the PCs. Therefore, the position of corrosion can be characterized by the feature image reconstructed by the PCs.

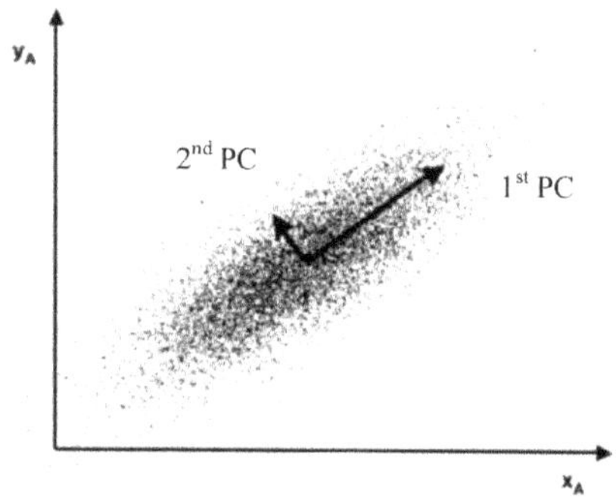

Figure 2. Schematic of feature extraction using PCA method [21].

2.3. Skewness and Kurtosis

For the whole process of thermal generation and heat dissipation, within the region where the induced eddy current exists, the surface temperature rises and decreases. The thermal image sequence can be regarded as the sampling for the temperature of the specimen's surface. Therefore, for a certain pixel P, the pixel value in every thermal image makes up a temperature variation curve, as shown in Figure 3.

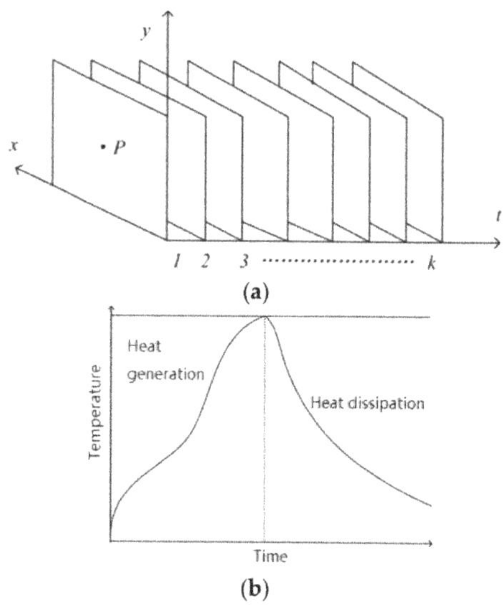

Figure 3. (**a**) Schematic of thermal image sequence, (**b**) temperature variation curve of pixel P.

Considering the temperature variation of each pixel is closely related to whether it is corroded or not, the curve should be investigated for corrosion characterization. Therefore, this paper extracts features from the shape of the temperature curves to indicate the surface state. The shape characteristic of the temperature curve can be expressed by two statistical features: skewness and kurtosis.

As shown in Figure 4, the skewness is greater than zero when the data are located to the right of the mean, i.e., negative skewness; less than zero when the data are concentrated on the left side, called positive skewness. When the data are in the normal distribution, the skewness is 0 [22]. The skewness is the ratio of third-order center distance to variance. The formula is as follows [23].

$$Skewness = \frac{n\sum(x_i - \overline{x})^3}{(n-1)(n-2)sd^3} \tag{7}$$

where n is the number of the data, x_i is the data value, $\overline{x}$ means the average value of data, sd represents the standard deviation.

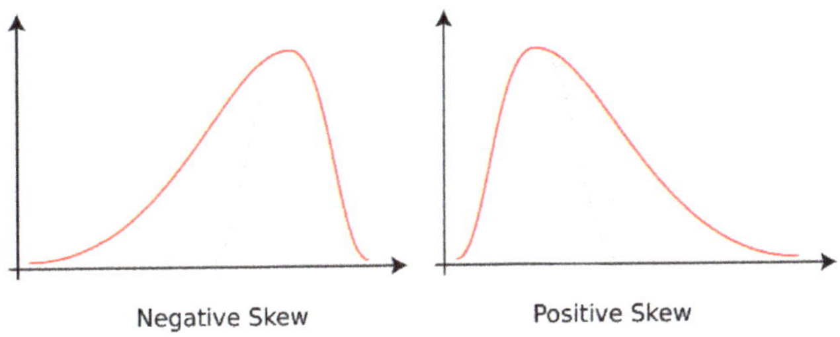

Figure 4. Schematic diagram of skewness. The blue dotted line means normal skewness [24].

The temperature variation is different between corroded and uncorroded regions. Therefore, the skewness value of the temperature curve at each point of the sample surface is calculated, and an image is reconstructed to indicate the region of corrosion. Similarly, kurtosis is used to describe the distribution of data. It is investigated in this paper for characterizing the paint-coated corrosion by extracting the feature of temperature variation.

As shown in Figure 5, positive kurtosis indicates that the data are distributed near the average value, and negative kurtosis shows that the data are more dispersed and the curve shape is shorter and fatter. In mathematics, kurtosis is the ratio of the fourth-order central moment to standard deviation, and the calculation formula is as follows [25]:

$$kurtosis = \frac{\sum(x_i - \overline{x})^4}{(n-1)sd^4} - 3 \tag{8}$$

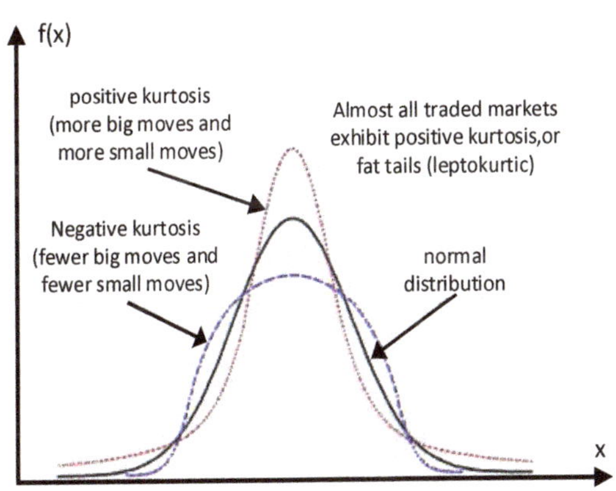

Figure 5. Schematic diagram of kurtosis.

In this paper, PCA, phase analysis, skewness and kurtosis are employed for thermal image sequence feature extraction. The extracted feature values are reconstructed into an image, which facilitates the recognition of corrosion, uncorroded, heating and non-heating regions, for subsequent corrosion evaluation.

3. Specimens and Experimental Setup

Low carbon steel S275, which is widely used in the engineering field, is selected for corrosion samples. The composition of S275 is listed in Table 1. Samples are prepared in the size of 300 × 150 × 3 (length × width × height) mm^3. These samples are covered by anticorrosive paint except the central area (30 × 30 mm^2) to keep the steel beneath clean and dry. Then, the samples are exposed to a marine atmosphere at Blyth UK for a certain time (1, 3, 6, 10 and 12 months). After a certain exposure time, local corrosion forms on the area without anticorrosive paint covered. Images of coated and uncoated rust patches are shown in Figure 6.

Table 1. The composition of S275 steel (wt%).

C	Si	Mn	Ni	S	P	Cr	N	Cu
<0.22	0.05–0.15	<0.65	<0.3	<0.05	<0.04	<0.3	0.012	<0.3

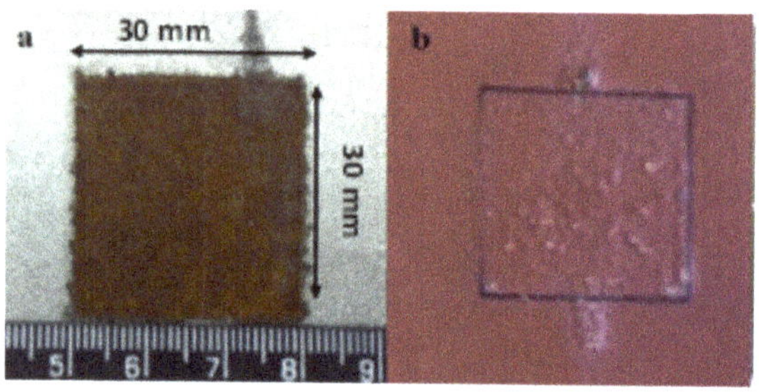

Figure 6. The images of (**a**) 1 month uncoated sample, (**b**) 1 month coated rust patch.

The experimental apparatus is shown in Figure 7. EasyHeat224 from Cheltenham Induction Heating provides $380A_{rms}$ AC excitation current. According to previous studies, the electrical conductivity and relative permeability of steel are 4.68 × 10^6 S/m and 60, while the corrosion products are close to 0.75 × 10^6 S/m and 4 [12]. For eddy current generation, a rectangular coil is made of a 6.35 mm hollow copper tube, which connects to a water pump for cooling. The excitation frequency is set to 260 khz to generate heat on the sample surface. Based on this frequency, the initial heated region is controlled within 60 μm. The excitation time is limited to 200 ms to avoid obvious temperature rise.

Figure 7. The experiment system.

A rectangular excitation coil is made of a 6.35 mm hollow copper tube, which is connected to a pump for water-cooling. An uncooled infrared camera with a 640 × 480 array of 17 μm detector, FLIR A655SC, from FLIR Systems, Inc., Wilsonville, OH, USA, is utilized to acquire IR image sequence with 50 hz sampling frequency and 30 mk temperature sensitivity. Considering the cooling process with no extra device, the whole record period is set to 2 s. A signal generator, Keysight 33500B, from Keysight Technologies Inc., Santa Rosa, CA, USA, is employed to generate a square pulse for synchronizing the excitation current and IR camera acquisition. The image sequence recorded by the infrared camera is transmitted to the PC, and then the four algorithms are performed and compared by MATLAB 2021b.

4. Results and Discussion

Firstly, the acquired infrared thermal imaging sequence is processed using the FFT algorithm with MATLAB. The reconstructed amplitude image and phase image are obtained, respectively. As shown in Figure 8, it is the amplitude image and phase image for 1, 3, 6, 10 and 12 months corroded samples using FFT processing. According to R. Yang's research, the amplitude image at 5 hz has the best performance in the amplitude images. However, it is difficult to characterize the corrosion shape clearly. The better effect in phase image is 4 hz, which can easily determine the shape of corrosion (the red area of the middle square is the corrosion area). Therefore, this paper will only analyze and discuss the phase frequency image of 4 hz to compare with the reconstructed image of other methods.

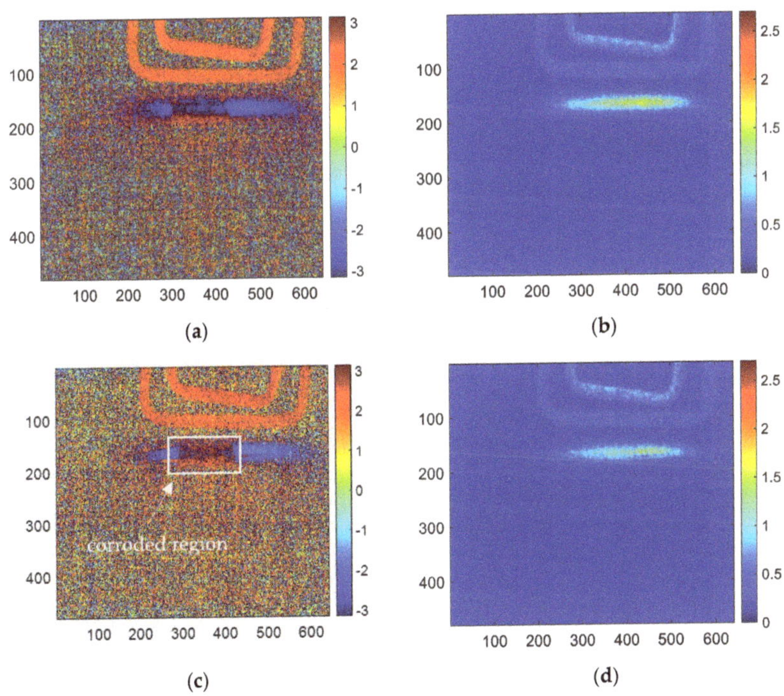

Figure 8. *Cont.*

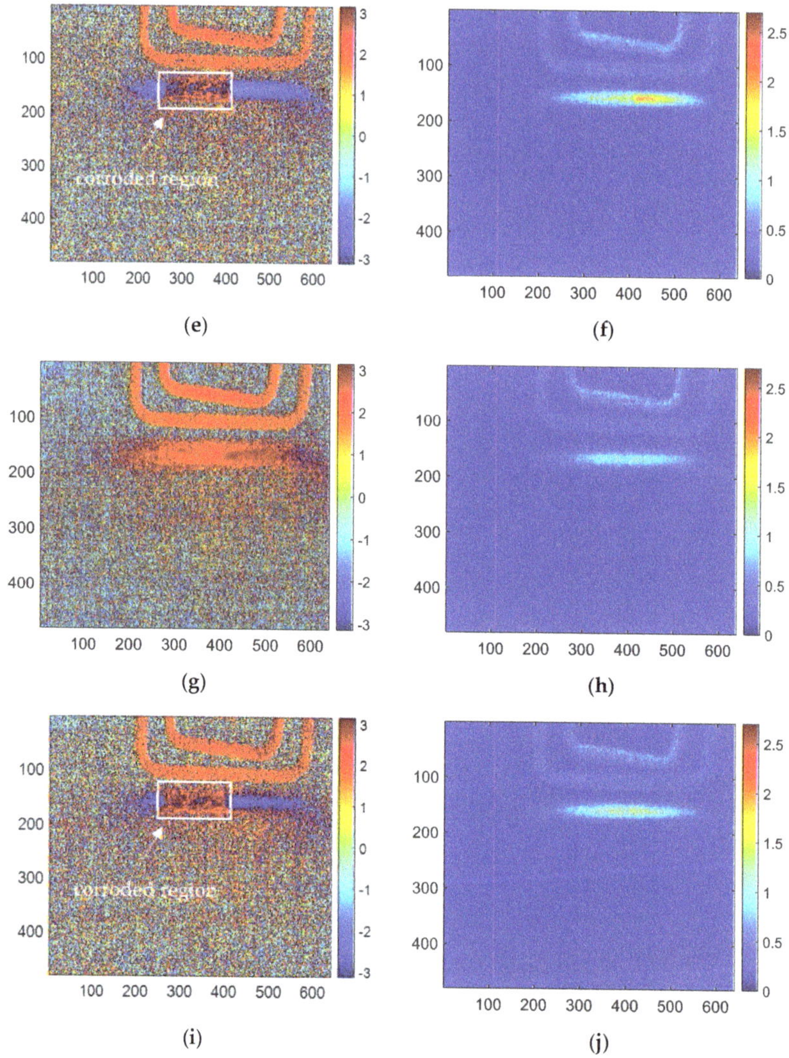

Figure 8. (**a**) 4 hz phase-frequency image of 1-month corrosion, (**b**) 5 hz amplitude-frequency image of 1-month corrosion, (**c**) 4 hz phase-frequency image of 3-month corrosion, (**d**) 5 hz amplitude-frequency image of 3-month corrosion, (**e**) 4 hz phase-frequency image of 6-month corrosion, (**f**) 5 hz amplitude-frequency image of 6-month corrosion, (**g**) 4 hz phase-frequency image of 10-month corrosion, (**h**) 5 hz amplitude-frequency image of 10-month corrosion, (**i**) 4 hz phase-frequency image of 12-month corrosion, (**j**) 5 hz amplitude-frequency image of 12-month corrosion.

For PCA methods, the sum of contribution rates of the first and second PCs is close to 90%, which means the first two PCs contain the most information of the original thermal image sequences. Therefore, only the first two PCs are analyzed and discussed. The reconstructed images of the first and second PCs for all the samples are shown in Figure 9. In this figure, (a), (c), (e), (g), (i) are the images of the first PC, which can hardly detect the corrosion area. However, (b), (d), (f), (h), and (j) are the reconstructed images of the second PC, which can clearly characterize the shape of the corrosion. Therefore, later in this paper, only the images of the second PC of PCA will be analyzed and discussed to compare with other methods.

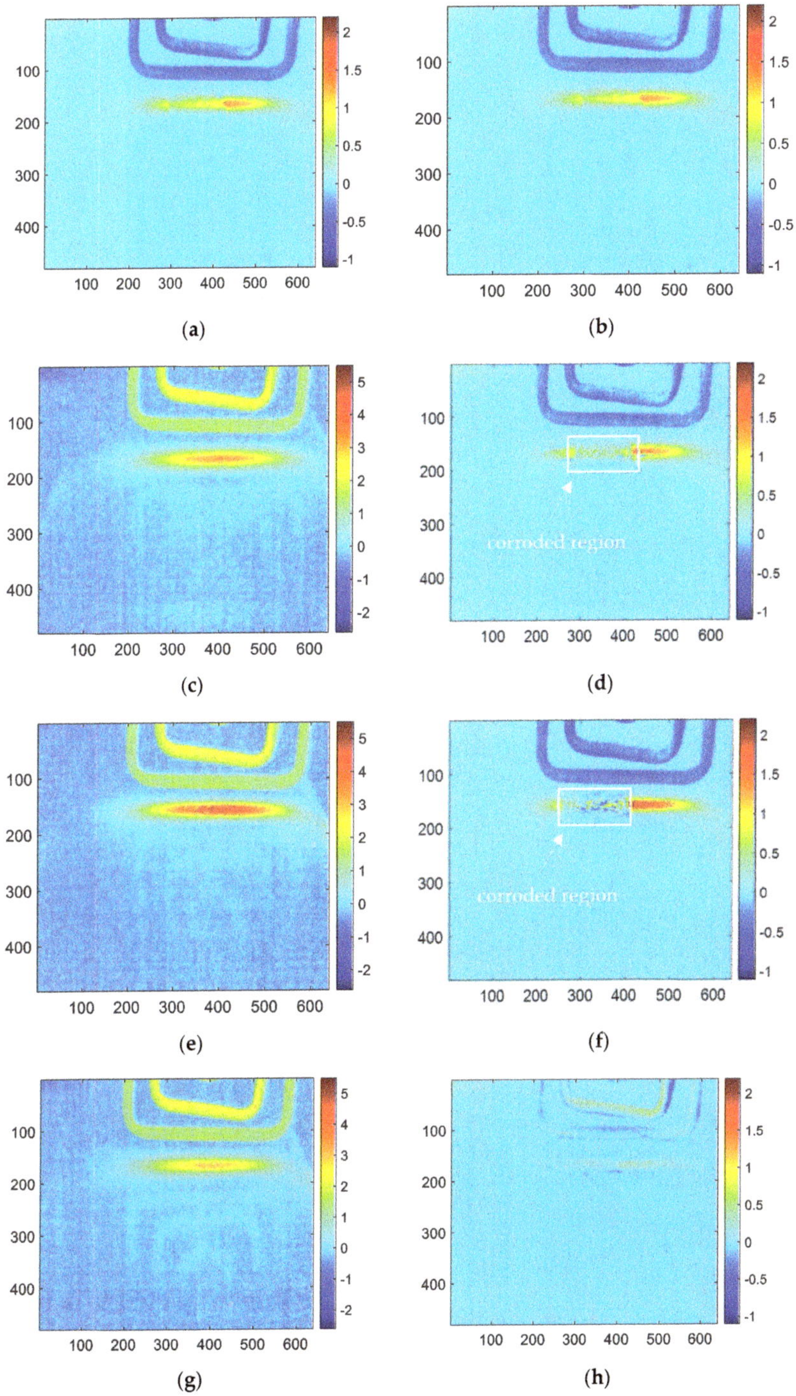

Figure 9. *Cont.*

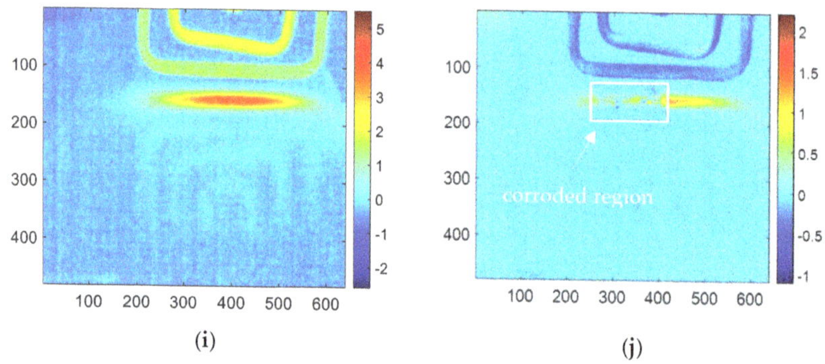

Figure 9. The reconstructed images of (**a**) first PC for 1-month corrosion, (**b**) second PC for 1-month corrosion, (**c**) first PC for 3-month corrosion, (**d**) second PC for 3-month corrosion, (**e**) first PC for 6-month corrosion, (**f**) second PC for 6-month corrosion, (**g**) first PC for 10-month corrosion, (**h**) second PC for 10-month corrosion, (**i**) first PC for 12-month corrosion, (**j**) second PC for 12-month corrosion.

The results of skewness analysis and kurtosis analysis based on temperature change curve analysis are shown in Figure 10. In this figure, (a), (c), (e), (g), and (i) are the corrosion images of 1, 3, 6, 10 and 12 months processed by the skewness method, respectively; (b), (d), (f), (h), and (j) are 1, 3, 6, 10 and 12 months corrosion images processed by kurtosis. The thermal imaging reconstitution processed by the skewness method can clearly characterize the corrosion shape, while the thermal imaging reconstitution processed by the kurtosis method cannot clearly characterize the corrosion. Based on the comparison of reconstructed images, this paper only discusses the images of the skewness method.

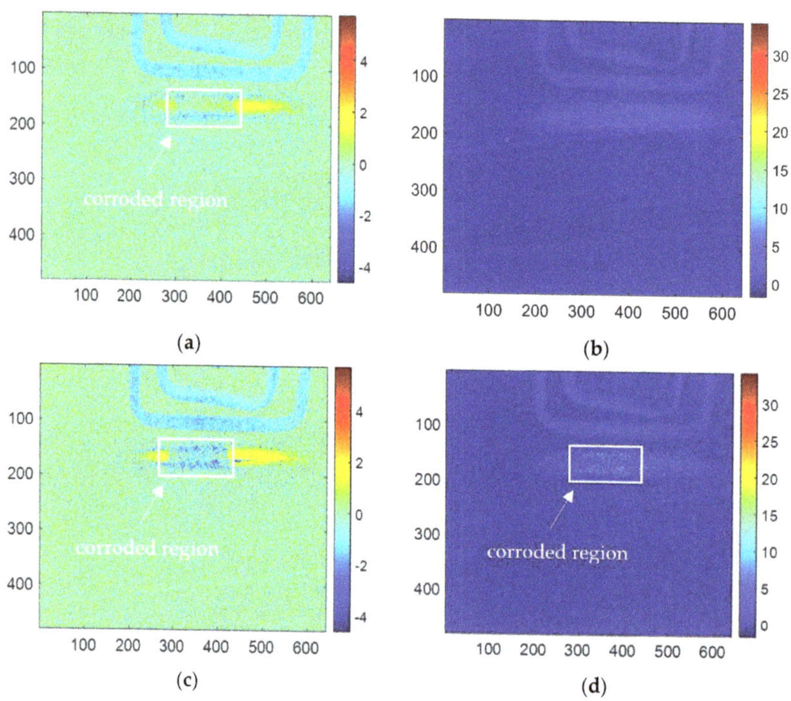

Figure 10. *Cont.*

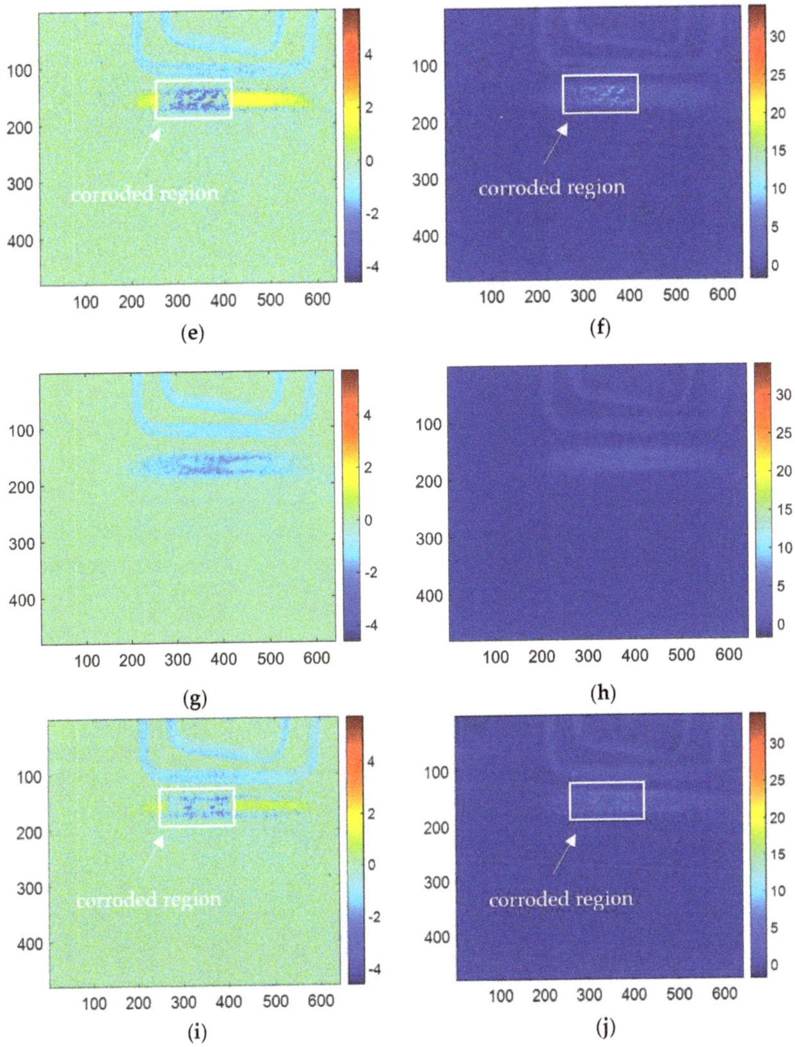

Figure 10. The reconstructed image of (**a**) skewness for 1-month corrosion, (**b**) kurtosis for 1-month corrosion, (**c**) skewness for 3-month corrosion, (**d**) kurtosis for 3-month corrosion, (**e**) skewness for 6-month corrosion, (**f**) kurtosis for 6-month corrosion, (**g**) skewness for 10-month corrosion, (**h**) kurtosis for 10-month corrosion, (**i**) skewness for 12-month corrosion, (**j**) kurtosis for 12-month corrosion.

Among the proposed methods, FFT, PCA, and skewness can identify the corrosion area by the extracted features. While kurtosis only can indicate a vague shape of the corroded region for 6-month and 10-month samples. Furthermore, in order to compare the performance of the features for corrosion evaluation, the phase, second PC, and skewness are selected for further investigation.

Considering the infrared thermal imaging data are interfered with by the instrument itself, environmental temperature and other factors, the features in the corroded region (region ① in Figure 11) and uncorroded region (region ② in Figure 11) are averaged, respectively, based on the reconstructed images. The square of region ① has 30 × 80 pixels. Region ② is the uncorroded region, which is processed as well as region ①. The averaged features are shown in Table 2. All of the averaged features for corroded and uncorroded regions are plotted in Figure 12a. Furthermore, the averaged features are normalized between 0 to 1 for comparison as shown in Figure 12b.

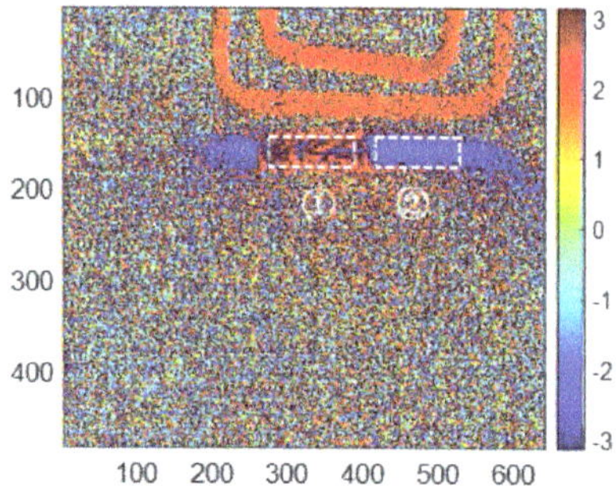

Figure 11. Phase frequency image of 4 Hz with FFT feature extraction. ① is the corroded region, ② is the uncorroded region.

Table 2. The averaged features of the corroded region.

The Mean Value of Feature of Corrosion Region	0 Months	1 Month	3 Months	6 Months	10 Months	12 Months
Phase (4 hz)	0.6060	0.1286	−0.1816	−0.2201	−0.6595	−0.1101
second PC	0.8128	0.5262	0.2867	0.1269	0.0922	0.4473
skewness	0.7283	0.0156	−0.9512	−1.4242	−1.4176	−0.6191

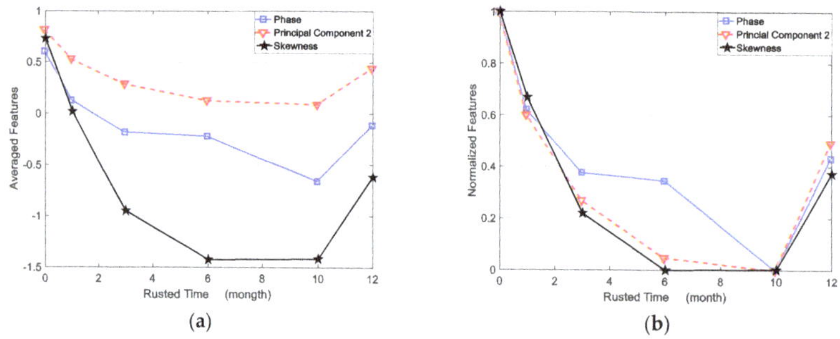

Figure 12. (**a**) Averaged features versus exposure time, (**b**) Normalized features versus exposure time.

From Figure 12a, the averaged second PC has the smallest variation from the uncorroded sample to the 10-month corroded sample, while the averaged skewness has the most obvious change for the same samples. That means the skewness feature has the best sensitivity for corrosion time evaluation, especially from 0 to 6 months of corroded time. As shown in Figure 12b, the normalized three features present non-monotonic variation versus corroded time. Averaged phase and second PC features decrease from an uncorroded state to 10 months corrosion, while increasing from 10 to 12 months corroded time. Averaged skewness decreases from 0 to 6 months of corroded time, but increases from 6 to 12 months of corroded time. Therefore, the selected three features can be used to evaluate the corrosion state before 6 months of corroding time. In particular, the averaged phase and second PC can recognize the corroded time till 10 months.

According to ref. [26], the induced heat power (P_h) is governed by:

$$P_h \propto I_e^2 \sqrt{\frac{\mu f}{\sigma}}, \quad \sigma = \frac{\sigma_0}{1+\alpha(T-T_0)} \tag{9}$$

where I_e is the excitation current, μ and f are the samples' magnetic permeability and excitation frequency, respectively. σ is the electrical conductivity, which is related to temperature T and T_0. α and σ_0 are the temperature coefficient of resistivity and the electrical conductivity at T_0, respectively. Considering the temperature variations of this experiment below is at most 0.72 °C, the Formula (9) can be simplified as $P_h \propto k\sqrt{\frac{\mu}{\sigma_0}}$. With the experimental setup, $k = I_e^2\sqrt{f}$ is a constant. Based on Formula (2) and $W = P_h \cdot t$, which generates the Joule heat, Formula (9) can be rewritten as Equation (10) in the same heating conditions.

$$T \propto k\sqrt{\frac{\mu}{\sigma_0}} \tag{10}$$

That means the magnetic permeability increases and electrical conductivity decreases leading to a temperature rise. For investigating the corrosion properties effect on the temperature variation, the average temperature of the uncorroded region (region ② in Figure 11) and corroded region (region ① in Figure 11) at 200 ms (the highest temperature) and 0 ms (non-heating temperature) are computed. Then, the maximum temperature variation can be calculated for comparison in Table 3.

Table 3. The maximum temperature variation of the uncorroded and corroded region.

Temperature Variation	1 Month	3 Months	6 Months	10 Months	12 Months
Corroded region	0.4920	0.4699	0.6850	0.4183	0.7162
Uncorroded region	0.4678	0.4383	0.7037	0.4126	0.7194

From Table 3, for 1 month and 3 months, the maximum temperature variations of the uncorroded region are below the corroded region. Considering the oxide corrosion model, ferrous oxide (FeO) occurred first on the surface, which is soon covered by a stable ferric oxide (Fe_2O_3) layer. This oxide layer has very low electrical conductivity and permeability at room temperature [14,27]. After that, a ferromagnetic layer with the composition Fe_3O_4 formed between the previous oxide layers, which has obviously high electrical conductivity and permanent magnetism [28]. The initial corrosion products, goethite $\alpha-$FeOOH and hematite $\alpha-Fe_2O_3$, have lower electrical conductivity than steel. Also, hematite shows weak ferromagnetism, and goethite is anti-ferromagnetic at room temperature [23,29]. According to Equation (10), in the early stage, the corroded region has a more obvious temperature increase. However, over time not only does the magnetic permeability and electrical conductivity change but also the corrosion height alters, as shown in Figure 13, there are complex heat distributions for long-term corrosion. Comparing the data in Table 3, 6 months corroded sample has a higher temperature variation of uncorroded region than that of corroded region. As shown in Figure 13d, the height of 6 months corrosion is about 110 μm. Due to the electrical conductivity and permeability of corrosion products are significantly smaller than steel [23], which leads to a deeper heat distribution and lower surface temperature. Comparing the temperature variation of 3 months and 6 months of corrosion, we believe that electrical conductivity decrease dominates the Joule heat generation in early corrosion.

However, for long-term corrosion, the situation is much more complex. Figure 13c presents a more complicated surface state than that of the early corrosion stage. Material is loose and flakes off making Joule heat generated, distributed, and conducted in a different layer, which means the infrared information can hardly characterize the corrosion's physical properties. That is why the previous research only investigated the early corrosion before 6 months of exposure time.

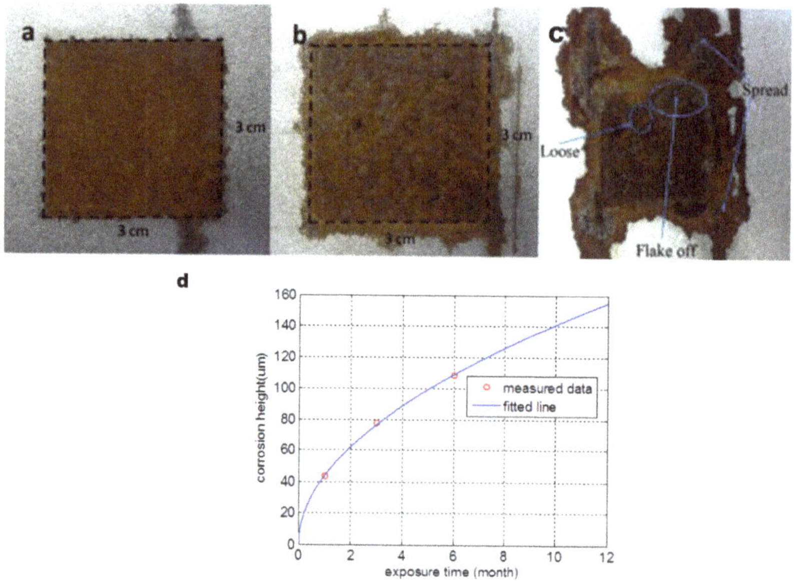

Figure 13. Image of corrosion state: (**a**) 1 month sample, (**b**) 6 months sample, (**c**) 10 months sample, and (**d**) heights versus exposure time [14].

5. Conclusions and Feature Work

In this paper, eddy current pulsed thermography is proven to be used to detect metal corrosion undercoating. Since the uneven heating and the emissivity anisotropy effect influence the thermal image acquisition, four feature extraction algorithms, FFT, PCA, skewness and kurtosis, are applied to compare the performance of feature extraction and corrosion evaluation. The main conclusion can be summarized as follows.

(1) Although the first principal component has more contribution, the reconstructed image of the second PC can reflect the characteristics of corrosion better.
(2) Despite the complex influence of corrosion properties and height, phase, skewness, and second PC can detect and recognize the early corrosion before 6 months. Especially, the averaged skewness has the most sensitivity among these corrosion evaluation approaches.
(3) The normalized second PC has the best performance for corrosion evaluation. It presents almost the same sensitivity as normalized skewness and a bigger measurement scale, 0 to 10 months corrosion.
(4) Based on the ECPT technique, all of these approaches can only evaluate the early-stage corrosion except PCA. Because the electrical conductivity dominates Joule heat generation and distribution (in depth) for early-stage corrosion. However, for long-term corrosion (more than 10 months exposure time), the height of corrosion means the infrared signals hardly characterize the physical properties of corrosion.

Considering the complex influence of physical properties variation on the sample surface infrared radiation, the proposed approach based on the ECPT technique can hardly characterize the long-time corrosion state. Some nondestructive testing methods that are based on a single physical principle, such as magnetic Barkhausen noise (MBN), and the eddy current test (ECT), can reflect the simple relations between corrosion state and magnetic or electrical properties. Therefore, with the advantages of data fusion and multi-parameter characterization, the ECPT technique combined with the MBN or ECT method can be the feature work for improving the corrosion state characterization and evaluation.

Author Contributions: Methodology, Y.Z.; Software, S.D.; Validation, S.D. and C.S.; Investigation, Y.W.; Resources, S.D.; Data curation, S.D. and C.S.; Writing—original draft, C.X.; Writing—review & editing, C.X., Y.Z. and Y.W. All authors have read and agreed to the published version of the manuscript.

Funding: China Construction Installation Group Co., Ltd. Project (AZ-2023-08), Key Laboratory of Nondestructive Testing (Fujian Polytechnic Normal University), Fujian Province University, National Key R&D Program of China (2020YFB1711203).

Institutional Review Board Statement: Not applicable.

Informed Consent Statement: Not applicable.

Data Availability Statement: The data used to support the findings of this study are included within the article.

Acknowledgments: This work is supported by Tian from Newcastle University, for the experiment guide and specimen preparation. The authors also wish to express their sincere gratitude to Shengqing Yao, who applied the financial support from China Construction Installation Group Co, Ltd., and checked the English writing.

Conflicts of Interest: The authors declare that there is no conflict of interest regarding the publication of this paper.

References

1. Morcillo, M.; Díaz, I.; Cano, H.; Chico, B.; de la Fuente, D. Atmospheric corrosion of weathering steels. Overview for engineers. Part I: Basic concepts. *Constr. Build. Mater.* **2019**, *213*, 723–737. [CrossRef]
2. Morcillo, M.; Díaz, I.; Cano, H.; Chico, B.; de la Fuente, D. Atmospheric corrosion of weathering steels. Overview for engineers. Part II: Testing, inspection, maintenance. *Constr. Build. Mater.* **2019**, *222*, 750–765. [CrossRef]
3. Bardin, I.V.; Bautin, V.A.; Gudoshnikov, S.A.; Ljubimov, B.Y.; Usov, N.A. Measurement of weak magnetic feld of corrosion current of isolated corrosion center. *AIP Adv.* **2015**, *5*, 017143. [CrossRef]
4. Zou, F.; Cegla, F.B. On quantitative corrosion rate monitoring with ultrasound. *J. Electroanal. Chem.* **2018**, *812*, 115–121. [CrossRef]
5. Hattori, M.; Nishikata, A.; Tsuru, T. EIS study on degradation of polymer-coated steel under ultraviolet radiation. *Corros. Sci.* **2010**, *52*, 2080–2087. [CrossRef]
6. Hren, M.; Kosec, T.; Legat, A. Characterization of stainless steel corrosion processes in mortar using various monitoring techniques. *Constr. Build. Mater.* **2019**, *221*, 604–613. [CrossRef]
7. Zhang, Y.; Liu, X.; Xiong, L.; Chen, Z.; Wei, J. A Review of NDT Methods for Wheel Burn Detection on Rails. *Sensors* **2023**, *23*, 5240. [CrossRef]
8. Li, K.; Tian, G.Y.; Cheng, L.; Yin, A.; Cao, W.; Crichton, S. State Detection of Bond Wires in IGBT Modules Using Eddy Current Pulsed Thermography. *IEEE Trans. Power Electron.* **2014**, *29*, 5000–5009. [CrossRef]
9. Zhu, J.Z.; Withers, P.J.; Wu, J.B.; Liu, F.; Yi, Q.J.; Wang, Z.J.; Tian, G.Y. Characterization of Rolling Contact Fatigue Cracks in Rails by Eddy Current Pulsed Thermography. *IEEE Trans. Ind. Inform.* **2021**, *17*, 2307–2315. [CrossRef]
10. Wang, Z.J.; Zhu, J.Z.; Tian, G.Y.; Ciampa, F. Comparative analysis of eddy current pulsed thermography and long pulse thermography for damage detection in metals and composites. *NDT E Int.* **2019**, *107*, 102155. [CrossRef]
11. Pan, M.; He, Y.Z.; Tian, G.Y.; Chen, D.X.; Luo, F.L. Defect characterisation using pulsed eddy current thermography under transmission mode and NDT applications. *NDT E Int.* **2012**, *52*, 28–36. [CrossRef]
12. Gotoh, Y.; Hirano, H.; Nakano, M.; Fujiwara, K.; Takahashi, N. Electromagnetic nondestructive testing of rust region in steel. *IEEE Trans. Magn.* **2005**, *41*, 3616–3618. [CrossRef]
13. He, Y.Z.; Tian, G.Y.; Pan, M.C.; Chen, D.X.; Zhang, H. An investigation into eddy current pulsed thermography for detection of corrosion blister. *Corros. Sci.* **2014**, *78*, 1–6. [CrossRef]
14. Yang, R.Z.; He, Y.Z.; Zhang, H.; Huang, S.D. Through coating imaging and nondestructive visualization evaluation of early marine corrosion using electromagnetic induction thermography. *Ocean Eng.* **2018**, *147*, 277–288. [CrossRef]
15. Yin, A.; Gao, B.; Yun Tian, G.; Woo, W.L.; Li, K. Physical interpretation and separation of eddy current pulsed thermography. *J. Appl. Phys.* **2013**, *113*, 064101. [CrossRef]
16. Xiong, Y.B.; Ye, C.; Zheng, M.Q.; Shi, J. Computed method and validation of thermal conductivity of austenitic stainless steel in cryogenic temperature range. *Cryog. Supercond.* **2018**, *46*, 19–23.
17. Takeda, M.; Onishi, T.; Nakakubo, S.; Fujimoto, S. Physical Properties of Iron-Oxide Scales on Si-Containing Steels at High Temperature. *Mater. Trans.* **2009**, *50*, 2242–2246. [CrossRef]
18. He, Y.Z.; Tian, G.Y.; Zhang, H.; Alamin, M.; Simm, A.; Jackson, P. Steel Corrosion Characterization Using Pulsed Eddy Current Systems. *IEEE Sens. J.* **2012**, *12*, 2113–2120. [CrossRef]
19. Maldague, X.; Marinetti, S. Pulse phase infrared thermography. *J. Appl. Phys.* **1996**, *79*, 2694–2698. [CrossRef]

20. Johnathan, H.; Quentin, F.; Khanh, V.; Seetha, R. Detection of corrosion under insulation on aerospace structures via pulsed eddy current thermography. *Aerosp. Sci. Technol.* **2022**, *121*, 107317.
21. Shlens, J. A tutorial on principal component analysis. *Int. J. Remote Sens.* **2014**, *51*, 2.
22. Harvey, C.; Siddique, A.R. Autoregressive conditional skewness. *J. Financ. Quant. Anal.* **1999**, *34*, 465–487. [CrossRef]
23. Ding, S.; Tian, G.Y.; Zhu, J.Z.; Chen, X.T.; Wang, Y.Q.; Chen, Y.M. Characterisation and evaluation of paint-coated marine corrosion in carbon steel using eddy current pulsed thermography. *NDT E Int.* **2022**, *130*, 102678. [CrossRef]
24. Skewness. Available online: https://encyclopedia.thefreedictionary.com/skewness (accessed on 1 May 2023).
25. Ahonen, T.; Hadid, A.; Pietikäinen, M. Face description with local binary patterns: Application to face recognition. *IEEE Trans. Pattern. Anal. Mach. Intell.* **2006**, *28*, 2037–2041. [CrossRef]
26. Liu, X.L.; Tian, G.Y.; Yu, C.; Luo, H.Z. Non-Contact Degradation Evaluation for GBT Modules Using Eddy Current Pulsed Thermography Approach. *Energies* **2020**, *13*, 2613. [CrossRef]
27. Canter, N. Tech beat: Conductivity in minerals. *Tribol. Lubr. Technol.* **2008**, *64*, 14–15.
28. Greenwood, N.N.; Earnshaw, A. *Chemistry of the Elements*, 2nd ed.; Butterworth-heinemann: Oxford, UK, 1997.
29. Guskos, N.; Papadopoulos, G.J.; Likodimos, V.; Patapis, S.; Yarmis, D.; Przepiera, A.; Przepiera, K.; Majszczyk, J.; Typek, J.; Wabia, M.; et al. Photoacoustic, EPR and electrical conductivity investigations of three synthetic mineral pigments: Hematite, goethite and magnetite. *Mater. Res. Bull.* **2002**, *37*, 1051–1061. [CrossRef]

sensors

MDPI

Article

Quantitative Visualization of Buried Defects in GFRP via Microwave Reflectometry

Ruonan Wang, Yang Fang *, Qianxiang Gao, Yong Li *, Xihan Yang and Zhenmao Chen

State Key Laboratory for Strength and Vibration of Mechanical Structures, Shaanxi Engineering Research Centre of NDT and Structure Integrity Evaluation, School of Aerospace Engineering, Xi'an Jiaotong University, Xi'an 710049, China; w2502209459@stu.xjtu.edu.cn (R.W.); gaoqianxiang@stu.xjtu.edu.cn (Q.G.); yxihan@stu.xjtu.edu.cn (X.Y.); chenzm@mail.xjtu.edu.cn (Z.C.)
* Correspondence: fangyang@xjtu.edu.cn (Y.F.); yong.li@mail.xjtu.edu.cn (Y.L.)

Abstract: Glass fiber-reinforced polymer (GFRP) is widely used in engineering fields involving aerospace, energy, transportation, etc. If internal buried defects occur due to hostile environments during fabrication and practical service, the structural integrity and safety of GFRP structures would be severely undermined. Therefore, it is indispensable to carry out effective quantitative nondestructive testing (NDT) of internal defects buried within GFRP structures. Along with the development of composite materials, microwave NDT is promising in non-intrusive inspection of defects in GFRPs. In this paper, quantitative screening of the subsurface impact damage and air void in a unidirectional GFRP via microwave reflectometry was intensively investigated. The influence of the microwave polarization direction with respect to the GFRP fiber direction on the reflection coefficient was investigated by using the equivalent relative permittivity calculated with theoretical analysis. Following this, a microwave NDT system was built up for further investigation regarding the imaging and quantitative evaluation of buried defects in GFRPs. A direct-wave suppression method based on singular-value decomposition was proposed to obtain high-quality defect images. The defect in-plane area was subsequently assessed via a proposed defect-edge identification method. The simulation and experimental results revealed that (1) the testing sensitivity to buried defects was the highest when the electric-field polarization direction is parallel to the GFRP fiber direction; and (2) the averaged evaluation accuracy regarding the in-plane area of the buried defect reached approximately 90% by applying the microwave reflectometry together with the proposed processing methods.

Keywords: electromagnetic nondestructive testing; microwave reflectometry; glass fiber-reinforced polymer; buried defects; quantitative evaluation

1. Introduction

Because of the excellent flexural strength, high stiffness-to-weight ratio, and outstanding corrosion resistance and thermal insulation [1–3], glass fiber-reinforced polymer (GFRP) is considered one of the preferable structural materials for fuselage skins, concrete reinforced beams, transportation pipelines, etc. [4–6]. Hitherto, it has been extensively utilized in structures of aerospace, automotive, marine, transportation, infrastructure, and other engineering fields [7,8]. However, it is inevitable that the GFRP structure is vulnerable to structural defects that undermine the stiffness as well as strength of structures, and pose a severe threat to the integrity of GFRP. Therefore, it is imperative to conduct nondestructive testing (NDT) of defects in GFRP structures in an effort to ensure the safety of engineering structures.

It is noteworthy that among typical defects of GFRP structures, the buried defect is deemed as one of the most critical flaws. This is because the buried defect normally conceals itself either within the structure body (manifesting itself as the void normally occurring during fabrication) or on the back surface of the structure (manifesting itself as the localized back surface material loss normally occurring during practical service due to the lateral impact [9]). The characteristics of the buried defect and GFRP make the traditional nondestructive testing techniques such as visual testing and eddy current testing, etc. formidable for quantitative visualization of defects. Although the testing methods including ultrasonic testing [10,11], infrared testing [12], and radiographic testing [13] are currently adopted for GFRP inspection, they could have technical drawbacks, particularly regarding the evaluation of the buried defect. Though air-coupled ultrasonic testing barely needs coupling [11], in the air, high-frequency signals decay rapidly, and thus the testing sensitivity to the buried defect could be low. The disadvantage of infrared testing lies in the fact that the accuracy in the detection of the buried defect could be limited due to the low thermal conductivity of GFRP. The major limitations of radiographic testing involve high costs, slow testing, and radiation hazards.

Compared with the aforementioned testing techniques, microwave testing has attracted much attention in the NDT field, especially regarding quantitative inspection of dielectric structures. The advantages of microwave testing include non-contact and remote testing, low attenuation in non-conductive materials, and one-side inspection of defects [14]. Besides, high-frequency electromagnetic waves with wide frequency bands are used in defect detection, making the testing capable of efficient imaging of hidden flaws with high longitudinal and transverse resolutions. As a result, microwave testing is found to have good detection accuracy for volumetric defects such as wall-thinning defects, and interlaminar flaws such as debonding and delamination in single-layer/multi-layer dielectric structures. Li et al. combined near-field and far-field microwave scanning imaging to obtain the full-size defect image of the large stiffened composite structure, greatly improving the detection efficiency [15]. Mei et al. developed an automatic detection system for full-size composite insulators and air-gap defects, carbonization defects, and conductive defects were located [16]. Wang et al. detected a phase change paraffin inclusion in honeycomb sandwich panels in the near field and accurately characterized the boundary of the phase change inclusion [17]. In addition, microwave testing can also be applicable to the detection of surface cracks in metallic structures. Yu et al. investigated the imaging and recognition of the boundary of the corrosion cracks in aluminum alloy structure via microwave testing [18]. Studies regarding microwave-testing-based GFRP evaluation have also been conducted. Rahman et al. compared the detection performance of K-band circular, K-band, and Ka-band rectangular waveguide for external and internal defects in GFRP pipe [19]. Shrifan et al. presented a novel technique based on k-means unsupervised machine learning with K-band rectangular waveguide to identify defects in GFRP [20]. In the study by Wang et al., a microwave equiphase frequency truncation method was proposed to detect and evaluate the thickness of kissing defects in GFRP laminates [21]. Sutthaweekul et al. combined principal component analysis and synthetic aperture radar to detect and characterize flat-bottom holes in coated GFRP pipes [22]. Shrifan et al. used maximal overlap discrete wavelet packet transform and the Bi-directional long short-term memory network approach to distinguish disbands in GFRP with rectangular waveguide operating from 18 to 26.5 GHz [23].

From the publications, it can be noticed that as one of the microwave antennae, the rectangular waveguide is often used for the detection of defects in GFRP. However, to the authors' knowledge, few studies involve the influence of the anisotropic characteristics of GFRP on the testing results. For a rectangular waveguide with TE_{10} mode as the fundamental mode of propagation, the direction of the electric field in the radiated microwave, i.e., polarization direction is parallel to the short edge of the waveguide aperture. When a unidirectional GFRP is inspected by the rectangular waveguide, the fiber direction of the GFRP has a certain angle concerning the wave-polarization direction. According to relevant research [24], when the included angle is different, the relative permittivity of GFRP in the detection system changes accordingly. This implies that (1) the anisotropic characteristics of GFRP influence the testing results; and (2) with the angle known, the so-called equivalent permittivity of GFRP, which can be computed with the corresponding formula mentioned in reference [25], could be applied for investigation regarding the influence of the wave-polarization direction with respect to the GFRP fiber direction on the reflection coefficient.

In this paper, the detection, imaging, and quantitative assessment of buried defects in a unidirectional GFRP slab via microwave reflectometry, i.e., one of the measurement methods in microwave testing, were intensively investigated. Two typical angles between the microwave polarization direction and the GFRP fiber direction, i.e., parallel and orthogonal, were taken into account. Numerical simulation models for microwave testing of buried defects were established in an attempt to scrutinize the testing signal response to the defect with the derived equivalent dielectric constant from the theoretical study. An experimental system of Ka-band microwave testing and screening of GFRP structures was built up. The testing signal and defect images for two angle scenarios were further analyzed. A direct-wave suppression method based on single-value sequence optimization was proposed for the enhancement of the defect image quality. Based on the defect image, the subsequent process involving image interpolation, feature enhancement, and defect-edge recognition is realized in an effort to evaluate the in-plane area of the buried defect.

2. Theoretical Analysis and Simulations

2.1. Theoretical Analysis of Microwave Response to a Dielectric Material

Microwave reflectometry is based on high-frequency electromagnetic waves which are emitted from microwave probes/antennas, and usually scan the specimens under inspection at the normal incident angle. During wave propagation, the incident wave is reflected at each heterogeneous interface. Since the reflected wave is closely associated with the material properties and sizing parameters, by analyzing the resulting testing signal the parameters of defects in the specimen can be extracted [26]. Since GFRP is non-conductive, it can be taken as one of the dielectric materials. Theoretical analysis is conducted to investigate the microwave response to a dielectric slab subject to the buried defect with a length and width appreciably larger than the size of the microwave antenna aperture. It is assumed that the microwave generated by an antenna is the transverse electric and magnetic field (TEM) plane wave that propagates along the z direction in the Fraunhofer region. The wave-propagation model is exhibited in Figure 1. Because the electric-field direction, magnetic-field direction, and wave-propagation direction are orthogonal to each other, based on Maxwell's equations, the expression regarding the x-component of the electric field of TEM wave propagating within a lossless non-conducting medium can be formulated as Equation (1) [26]:

$$E_{xi}(z) = E_{xi}^{+}e^{-jkz} + E_{xi}^{-}e^{jkz}, k_i = \omega\sqrt{\mu_i\varepsilon_i} \tag{1}$$

where, the subscript i represents different dielectric material domains, and E_x^+ and E_x^- denote the electric-field intensities regarding the incident and reflected microwaves at the boundary of $z = 0$, respectively. k and ω stand for the wave number and angular frequency, respectively. μ is the permeability of the medium, which is essentially the product of the vacuum permeability and medium relative permeability. ε denotes the medium permittivity which is the product of the vacuum permittivity and relative permittivity of the medium. It is noted that for a plane TEM wave the y and z components of the electric field vanish. In regard to the magnetic field of the TEM wave, its y component can be written as Equation (2) [26]:

$$H_{yi}^{+}(z) = \frac{1}{\eta_i}E_{xi}^{+}(z), \eta_i = \sqrt{\frac{\mu_i}{\varepsilon_i}} \tag{2}$$

where, η is the intrinsic impedance of the medium. Note that the x and z components of the magnetic field vanish.

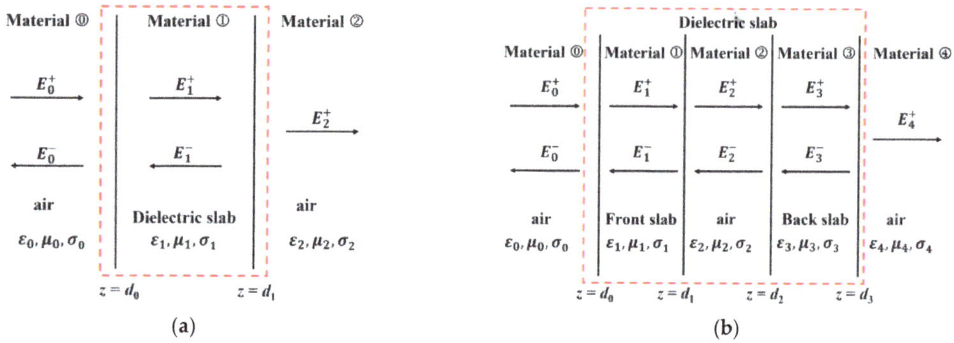

Figure 1. Electromagnetic wave propagation for microwave detection of buried defects: (**a**) electromagnetic wave propagation for the scenario with the back surface material loss; (**b**) electromagnetic wave propagation for the case with the internal hole.

Based on the model, the buried defects involving the back surface material loss and internal hole (simulating the void) can be modeled by either varying the thickness of the slab, i.e., d or introducing a planar air layer (with the thickness of $d_1 - d_2$) to the slab. When the slab is tested by the incident microwave, for the scenario of the back surface material loss in Figure 1a, based on Equations (1) and (2) in conjunction with the continuity of the electromagnetic field at $z = 0$ and $z = d_1$, the intensity of the electric field of the reflected wave E_0^- can be formulated as Equation (3):

$$\begin{cases} E_0^- = cE_0^+ \\ c = -\frac{j\left(\eta_2{}^2 - \eta_1{}^2\right)\tan(k_1 d_1)}{2\eta_1\eta_2 + j\left(\eta_1{}^2 + \eta_2{}^2\right)\tan(k_1 d_1)} \end{cases} \tag{3}$$

where, c denotes the reflection coefficient of the slab subject to the back surface material loss. η_1 and η_2 are the natural impedances of the air and flawless slab, respectively. k_1 is the wave number in the slab domain. It can be noticed from Equation (3) that the reflection coefficient c is highly associated with the variation in the slab thickness. By analyzing the reflection coefficient which is usually measured in microwave testing, the back surface material loss could be detected and evaluated.

For the scenario regarding the internal hole in Figure 1b, the relation of the reflected wave with the incident wave can be expressed in the matrix notation as Equation (4):

$$\begin{bmatrix} E_0^+ \\ E_0^- \end{bmatrix} = \prod_{t=1}^{3}\left\{\begin{bmatrix} \left(\frac{1}{2} + \frac{\eta_{t-1}}{2\eta_t}\right)e^{jd_{t-1}(-k_t + k_{t-1})} & \left(\frac{1}{2} - \frac{\eta_{t-1}}{2\eta_t}\right)e^{jd_{t-1}(k_t + k_{t-1})} \\ \left(\frac{1}{2} - \frac{\eta_{t-1}}{2\eta_t}\right)e^{jd_{t-1}(-k_t - k_{t-1})} & \left(\frac{1}{2} + \frac{\eta_{t-1}}{2\eta_t}\right)e^{jd_{t-1}(k_t - k_{t-1})} \end{bmatrix}\right\} \cdot \begin{bmatrix} \left(\frac{1}{2} + \frac{\eta_3}{2\eta_4}\right)e^{jd_3(-k_4 + k_3)} \\ \left(\frac{1}{2} - \frac{\eta_3}{2\eta_4}\right)e^{jd_3(-k_4 - k_3)} \end{bmatrix} E_4^+ = \begin{bmatrix} \alpha \\ \beta \end{bmatrix} E_4^+ \tag{4}$$

Based on Equation (4), the reflection coefficient of the slab subject to the internal hole can be readily computed by using the expression written as Equation (5):

$$c_2 = \frac{\beta}{\alpha} \tag{5}$$

It can be found from Equations (4) and (5) that c_2 relies highly on the material properties of the slab and sizing parameters of the defect. Consequently, the internal hole could be detected and assessed by analyzing c_2.

Both Equations (3) and (5) imply that the large-sized buried defect can be detected by analyzing the reflection coefficient which is correlated with the sizing parameters and electromagnetic properties of the slab. Apart from the defect size, special attention has been paid to the relative permittivity ε_r of the dielectric slab. It is noteworthy that Equations (1)–(5) hold merely under the condition that the slab is isotropic. They are inapplicable for anisotropic materials such as GFRP. However, the research by Kharkovsky et al. indicated that the apparent relative permittivity of GFRP depends on the wave-polarization angle between the microwave polarization direction of the incident wave and the fiber direction [24]. With the angle known in the testing setup, GFRP shows dielectric characteristics analogous to those of an isotropic dielectric material with the equivalent relative permittivity. This opens up the possibility of investigating the microwave response to a unidirectional GFRP with a buried defect since the anisotropic properties of GFRP can be characterized by various equivalent relative permittivity for different wave-polarization angles.

2.2. The Equivalent Relative Permittivity of GFRP and Simulations

Two wave-polarization angles are involved in the simulations: (1) the microwave polarization direction parallel to the fiber direction of a unidirectional GFRP slab; and (2) the microwave polarization direction orthogonal to the fiber direction. The schematic illustration is presented in Figure 2.

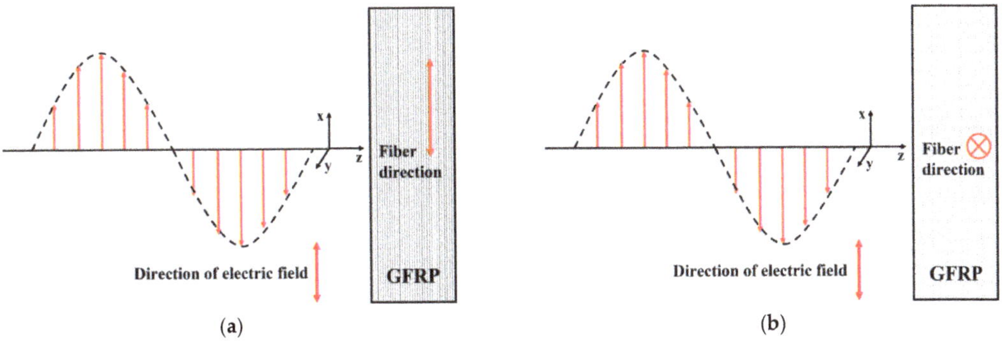

Figure 2. Two wave-polarization angles are involved in simulations: (**a**) the electric field direction parallel to the fiber direction; and (**b**) the electric field direction orthogonal to the fiber direction.

Based on [24], for each wave-polarization-angle case, the equivalent relative dielectric can be calculated by Wiener upper limit and Hashin Shtrikman upper bound Equations (6) and (7):

$$\varepsilon_{\parallel} = v_f \varepsilon_f + \left(1 - v_f\right)\varepsilon_m \tag{6}$$

$$\varepsilon_{\perp} = \varepsilon_f \left[1 + \frac{3\left(1 - v_f\right)\left(\varepsilon_f - \varepsilon_m\right)}{\left(\varepsilon_f - \varepsilon_m\right)v_f - 3\varepsilon_f}\right] \tag{7}$$

where, $\varepsilon_{\parallel}$ and $\varepsilon_{\perp}$ are the equivalent relative permittivity when the GFRP is illuminated by the incident microwave with the polarization direction parallel and orthogonal to the fiber direction, respectively. ε_f and ε_m denote the relative permittivity of the fiber and resin, which are associated with the volume fraction v_f regarding the glass fibers in the GFRP. The microwave which is reflected from the GFRP and sensed by an antenna is dependent on $\varepsilon_{\parallel}$ and $\varepsilon_{\perp}$. Therefore, the testing signal, aka., S_{11} acquired by a vector network analyzer (VNA) in microwave testing varies for the wave-polarization angles. It is noted that S_{11} is formulated as Equation (8):

$$S_{11} = \frac{\Gamma\left(1 - e^{2jkD}\right)}{1 - \Gamma^2 e^{2jkD}}, \Gamma = \frac{Z_S - 1}{Z_S + 1}, Z_S = \sqrt{\frac{\mu}{\varepsilon\prime}} \tag{8}$$

where, D denotes the thickness of the GFRP slab. Γ is the reflection coefficient at the heterogeneous interface. Z_s stands for the wave impedance in a function of magnetic permeability μ, $\mu = 4\pi \times 10^{-7}$ T·m/A and dielectric constant ε' which is either $\varepsilon_{\parallel}$ or $\varepsilon_{\perp}$ for each wave-polarization angle.

In reference to [27], the relative permittivity of the fiber is 6.2 whilst the relative permittivity of the resin is 3.0. The volume fraction of the fibers is normally 60% [27]. By using Equations (6) and (7), the equivalent relative permittivity of a typical unidirectional GFRP for two wave-polarization angles is $\varepsilon_{\parallel} = 4.92$ and $\varepsilon_{\perp} = 4.77$.

With the equivalent relative permittivity computed out, further simulations were conducted in an attempt to investigate the testing signal response to the localized buried defects inspected by a Ka-band rectangular waveguide with an aperture size of 7.11 mm × 3.56 mm. In consideration of the fact that the fundamental mode of the microwave excited by waveguide is TE_{10} and only y-component of the electric field exists and its intensity is independent of y coordinate, a 2D finite element model was built up. The schematic illustration of the model is shown in Figure 3 whilst the simulation parameters are listed in Table 1.

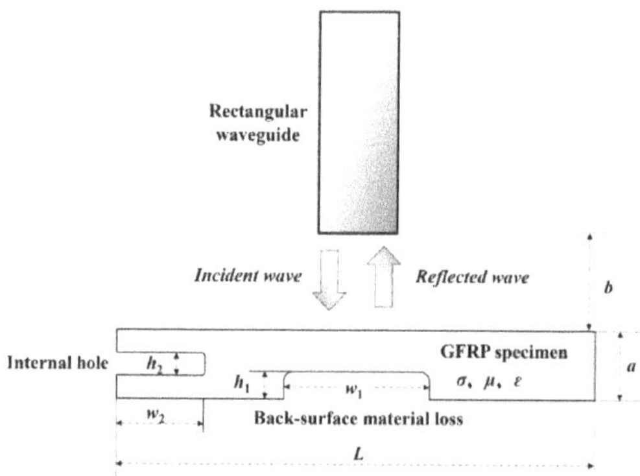

Figure 3. Schematic illustration of the simulation model regarding the microwave testing of a GFRP slab subject to buried defects.

Table 1. The modeling parameters.

Frequency Range	Stand-off Distance b	Specimen Size $L \times a$	Size of the Back-Surface Wall-Thinning Defect $w_1 \times h_1$	Size of the Internal Hole $w_2 \times h_2$
26.5 GHz~40 GHz	1 mm	200 mm × 6 mm	10 mm ×3 mm	20 mm × 2 mm

In simulations, the back surface material loss and the internal hole were individually tested by the rectangular waveguide deployed right over their centers. The number of the frequency samples was set as 136. After S_{11}, each frequency sample was derived and its mean amplitude ($\overline{S_{11}}$) was subsequently calculated. The simulated testing signals, i.e., S_{11} in the function of the excitation frequency are presented in Figure 4, whilst the mean amplitudes of S_{11} are tabulated in Table 2.

Table 2. The mean amplitude of S_{11} vs. the equivalent relative permittivity of the GFRP slab.

Equivalent Relative Permittivity	Flawless Region	Back-Surface Material Loss	Internal Hole
$\varepsilon_{\parallel} = 4.92$	−6.23 dB	−3.89 dB	−12.41 dB
$\varepsilon_{\perp} = 4.77$	−6.47 dB	−4.10 dB	−12.64 dB

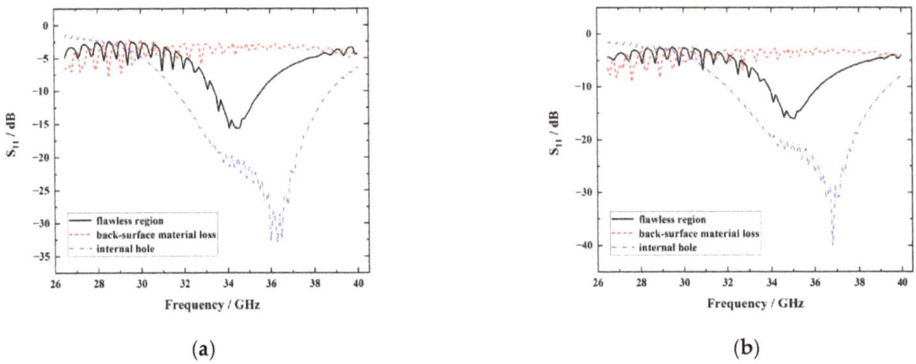

Figure 4. Simulated signals: (**a**) parallel; (**b**) orthogonal.

It can be seen from Figure 4 that there are obvious differences in the simulation signals at the flawless region, the back surface material loss, and the internal hole. Specifically, the S_{11} amplitude of the back surface material loss was the largest, followed by the flawless region while that of the internal hole was the smallest. It can also be seen from Table 2 that the mean amplitude of S_{11} for the back surface material loss was larger than that of the flawless region, whilst for the case of the internal hole it was the smallest. This is because compared with the flawless region, the back surface material loss reduced the loss of the incident wave as the thickness of the GFRP slab decreased, while the internal hole caused a lot of microwave scattering due to multiple interfaces. The results also imply that in the presence of buried defects, the mean amplitude of S_{11} with the microwave polarization direction parallel to the fiber direction is larger than that with the orthogonal scenario. Further analysis regarding the variation of the mean amplitude of S_{11} in the presence of buried defects reveals that it is bigger for the parallel scenario, indicating that the testing sensitivity is higher when the microwave polarization direction is parallel to the fiber direction.

3. Experiments for Quantitative Screening of Buried Defects

3.1. Experimental System Setup

In parallel to the simulations, experiments were conducted for investigation regarding quantitative screening of buried defects in GFRP via microwave reflectometry. An experimental system of microwave testing is hereby built up. The schematic illustration of the system is portrayed in Figure 5. The system consists of a VNA (Keysight N5224A) connected with an open-ended rectangular waveguide of Ka band, three-axis scanning table, scanner controller, and PC. The Ka-band incident microwave generated by the VNA is transmitted through the coaxial cable to the waveguide, which converts the microwave from the TEM mode into TE mode. During experiments, the waveguide was deployed over the specimen with the normal wave-incident setup. It also captures the reflected wave from the specimen simultaneously. The testing signal, i.e., S_{11} was subsequently obtained from the VNA. The experimental parameters are listed in Table 3.

Table 3. Parameters of the experimental setup.

Frequency Range	Frequency Samples	Scanning Interval	Stand-off Distance
26.5 GHz~40 GHz	1351	2 mm	1 mm

A planar GFRP sample containing epoxy resin and glass fiber was employed in experiments. It was fabricated with the automated fiber placement process. All the fibers in the GFRP sample were unidirectional. The sample size was 300 mm × 300 mm × 6 mm (length × width × thickness). The buried defects including the simulated back-surface material loss and internal hole were fabricated by machine work. The schematic illustration of the sample and the defects alongside their sizing parameters are exhibited in Figure 6. The back surface material loss includes two shapes: square and letter, simulating the defect with regular and irregular shapes, respectively. The internal hole was used to imitate voids in GFRP. All defect numbers and sizes are also shown in Figure 6, where *a* is the length of the square side, *h* is the depth of the material loss, *d* is the diameter of the internal hole, and *s* is the depth of the hole.

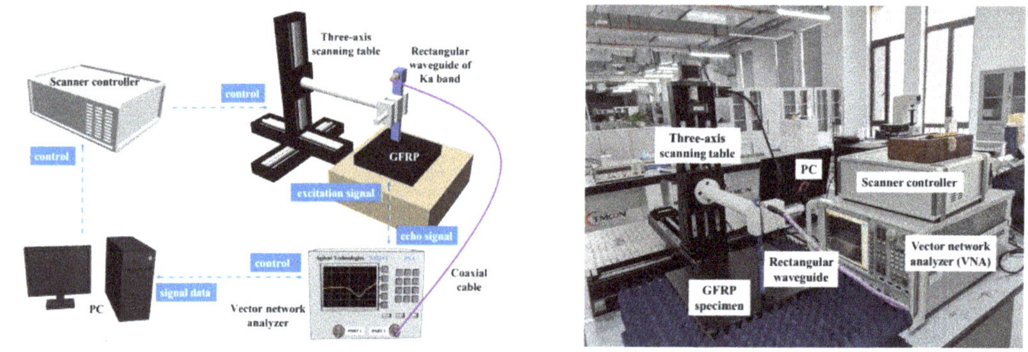

(a) (b)

Figure 5. Schematic illustration of the experimental system: (**a**) the diagram; (**b**) the practical picture.

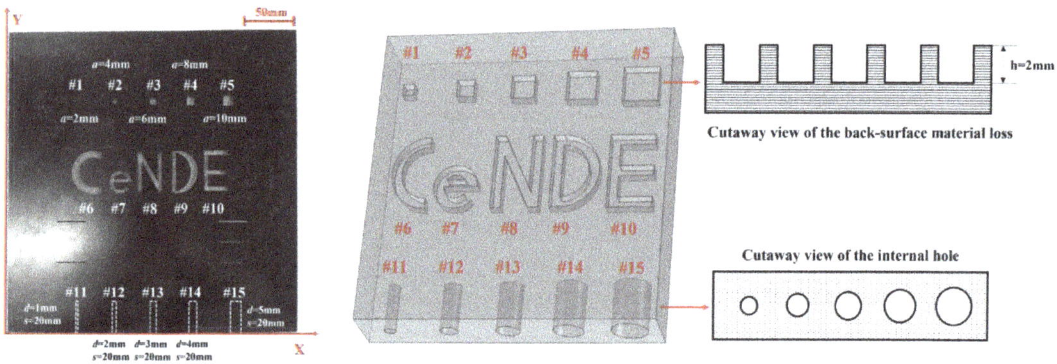

(a) (b)

Figure 6. The testing specimen: (**a**) the specimen picture; (**b**) the schematic illustration of the specimen.

To investigate the influence of the angles between the wave-polarization direction and the fiber direction on the detection results, the rectangular waveguide was placed over the specimen with the wave-polarization angle set as either 0° (the parallel case) or 90° (the orthogonal case), which is indicated in Figure 2.

3.2. Defect Testing and Imaging at Different Wave-Polarization Angles

In regard to two wave-polarization angles, testing signals obtained for the flawless region and the center of the back-surface material loss are shown in Figure 7a,c. For two-angle scenarios, the amplitude of S_{11} measured at the flawless region was lower than that at the defect center in the band of 26.5 GHz~31 GHz, indicating that testing signals are responsive to the presence of buried defects. The inverse Fourier transform is exploited on frequency–domain signals in Figure 7a,c to obtain time–domain signals, as shown in Figure 7b,d. It can be observed that with both the parallel and orthogonal scenarios, the first echo of time–domain signals at the defect center and the flawless region basically overlapped whilst the second echo exhibited differences according to zoom-in plots in Figure 7b,d. The second echo of the defect center was slightly higher than that of the flawless region, indicating that the energy of the reflected wave at the defect center is greater than that of the flawless region on account of the less loss on the incident wave at the center of the back surface material loss. It is noteworthy that the amplitude of S_{11} with the wave-polarization direction parallel to the fiber direction at the flawless region and the defect center was higher than that with of orthogonal case, which is consistent with the simulation results.

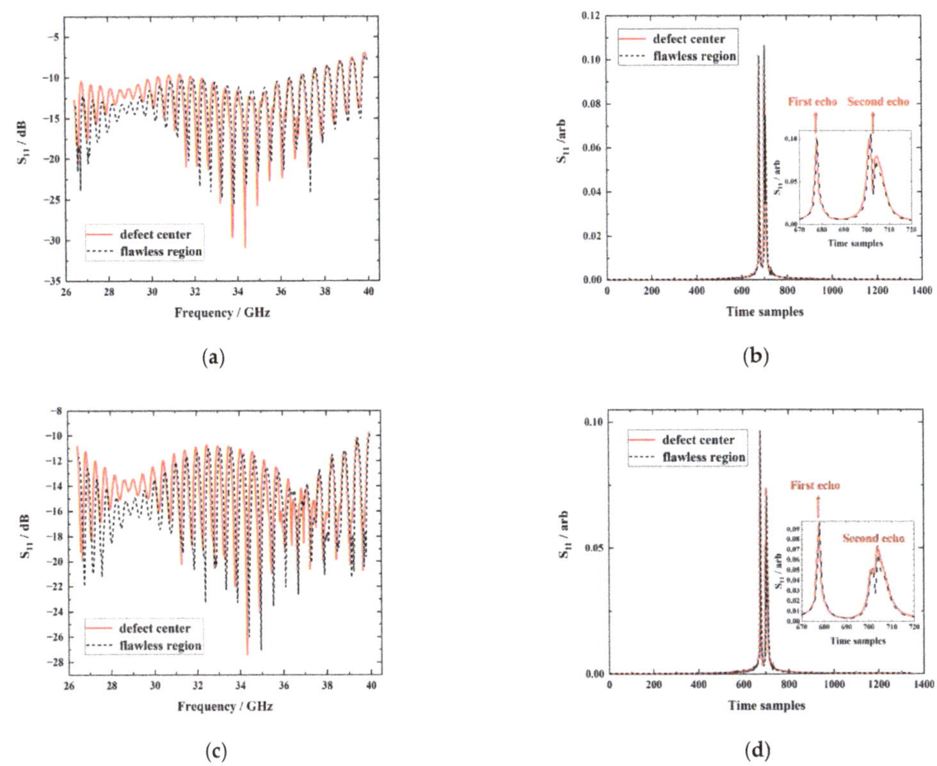

Figure 7. The testing signal for the flawless region and the center of the back surface material loss: (**a**) frequency–domain signals with the parallel case; (**b**) time–domain signals with the parallel case; (**c**) frequency–domain signals with the orthogonal case; (**d**) time–domain signals with the orthogonal case.

With the 2D scanning of the specimen, after testing signals are acquired at each scanning point, the area enclosed by the time–domain signal and the time–samples axis was extracted as the signal feature, which was denoted as SA and used for defect imaging. The defect image with SA as the pixel value at every scanning point is shown in Figure 8.

It can be observed from Figure 8 that although the profile of the back-surface material loss and the internal hole can be roughly observed in the figure, the contrast between the defect region and the flawless region is low. Referring to the results presented in Figure 7, it can be found that (1) there were overlapping sections in the time–domain signals for the flawless region and the defect center; and (2) the response of the second echo to the defect was barely significant. This implies that there could be many similar magnitude-based features in the experimental signals for the defect and the flawless region. Therefore, in order to enhance the testing sensitivity and thus image contrast between the defect region and the flawless region, it is indispensable to mitigate these similar magnitude-based features within testing signals.

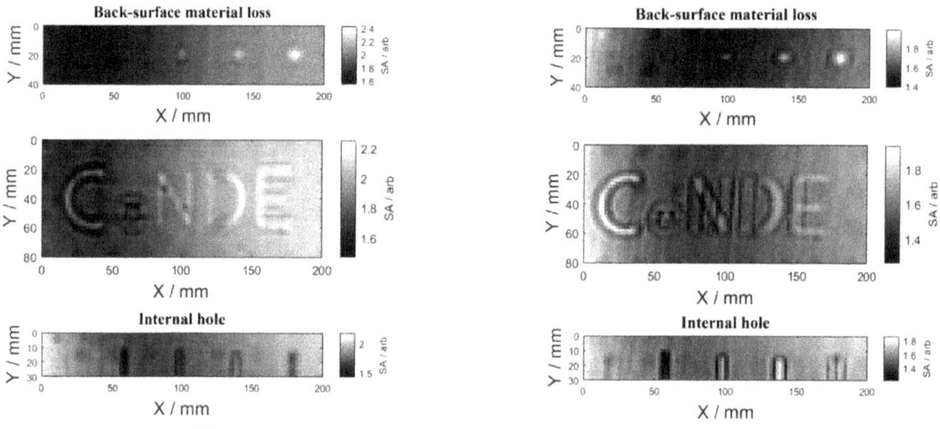

Figure 8. Defect images with the (**a**) parallel case and (**b**) orthogonal case.

3.3. Proposition of the Direct-Wave Suppression Method

In microwave testing, the raw testing signals received by the microwave probe normally consist of (1) the reflected signal conceiving the defect information; and (2) the direct wave signal due to the coupling at the port of the rectangular waveguide along with the reflected wave from the specimen surface. Apparently, the direct wave barely conceives the information on the buried defects in the GFRP specimen. Since the magnitude of the direct wave is considerably larger than that of the reflected signal, the defect information could be masked by the redundant information of the direct wave. In other words, the testing sensitivity can be enhanced by suppressing the direct wave.

In an effort to mitigate the influence of the direct wave on the testing results, particularly the defect images, a direct-wave suppression method based on single-value decomposition (SVD) is proposed. The general idea of this method is to obtain a diagonal matrix that stores single values of original reflected signals, in which the single values representing the direct-wave components are identified and subsequently set to 0. This is followed by the reflected signal being reconstructed by utilizing the diagonal matrix partially set to zero in conjunction with the inverse transformation of SVD. Specifically, the data matrix x_i^k (superscript denoting frequencies and subscript standing for scanning points) is assembled into the data matrix **A** according to Figure 9 with each row in the data matrix representing reflected signals from all scanning points at the same frequency and each column representing reflected signals from all frequencies at the same scanning point. Next, single-value decomposition of the data matrix **A** is performed according to Equation (9):

$$\mathbf{A} = \mathbf{U}\mathbf{S}\mathbf{V}^{\mathrm{T}} \tag{9}$$

where, **A** is the data matrix of the reflected signals, **U** is the left singular matrix, **V** is the right singular matrix, and **S** is the diagonal matrix with the elements on the diagonal standing for the eigenvalues of **A**, also known as singular values, arranged in descending order on the diagonal. Since according to the above analysis and relevant references the direct-wave component in the reflected signal was the strongest, the largest singular values in **S** were considered as the direct wave components and set to 0 [28]. The number of singular values set to 0 was determined by the contrast ratio of the defect image. The attempts to set the first certain number of singular values were repeated and the contrast ratio of the defect image obtained from various results of attempts was calculated. It was found that for the GFRP specimen with buried defects in this paper, when the first three largest singular values were set to 0, the contrast ratio of the defect image was highest. Hence, the direct wave components were determined to be the first three largest singular values. Ultimately, the reflected signal $\mathbf{A}'$ after suppressing the direct wave is calculated by putting the matrix $\mathbf{S}'$ partially set to zero, the matrix $\mathbf{U}$, and the matrix $\mathbf{V}$ into Equation (9). Accordingly, testing signals corresponding to Figure 7, after the direct waves' suppression, are shown in Figure 10.

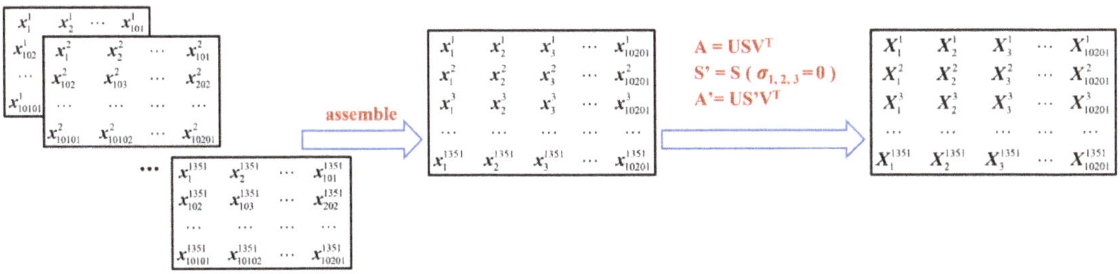

Figure 9. The direct wave suppression method based on SVD.

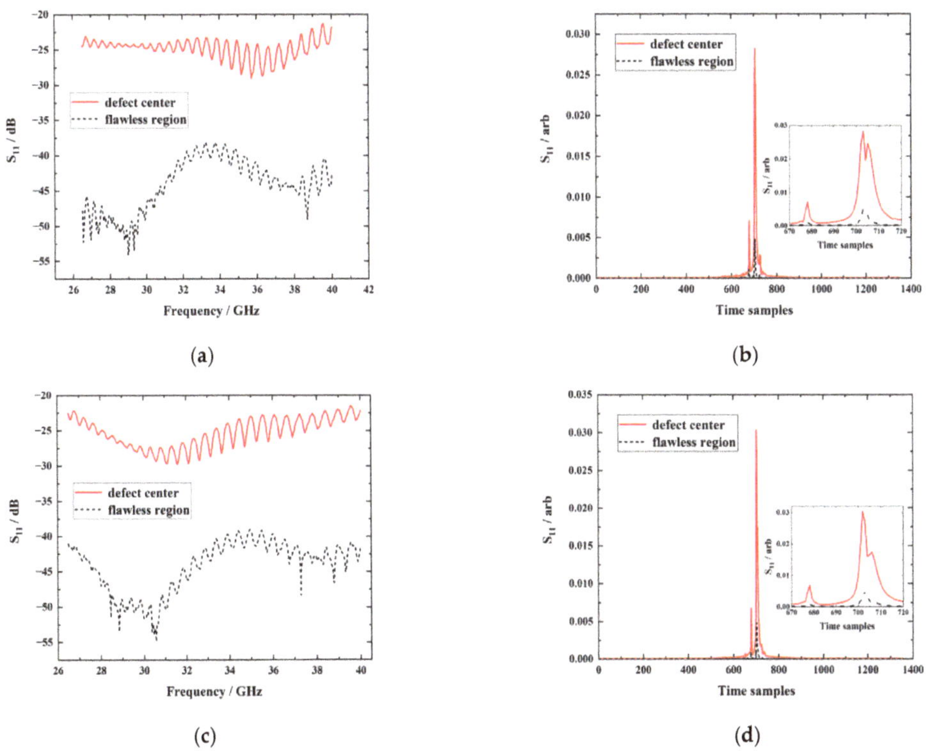

Figure 10. The testing signal after direct wave suppression: (**a**) frequency–domain signals with the parallel case; (**b**) time–domain signals with the parallel case; (**c**) frequency–domain signals with the orthogonal case; (**d**) time–domain signals with the orthogonal case.

It can be seen from the figure that after the direct wave suppression by the proposed SVD method, the frequency–domain signal and the time–domain signal at the defect center and the flawless region showed significant differences in that both the frequency–domain and time–domain signals of the flawless region were distinctly below that of the defect center. Comparing Figure 10 with Figure 7, it is noticeable that the signal difference between the defect center and the flawless region was significantly enhanced, implying the applicability of the proposed method for direct wave suppression. SA is still extracted from the reflected signal after the direct wave suppression and used for reproduction of the defect images. The reproduced defect images are shown in Figure 11.

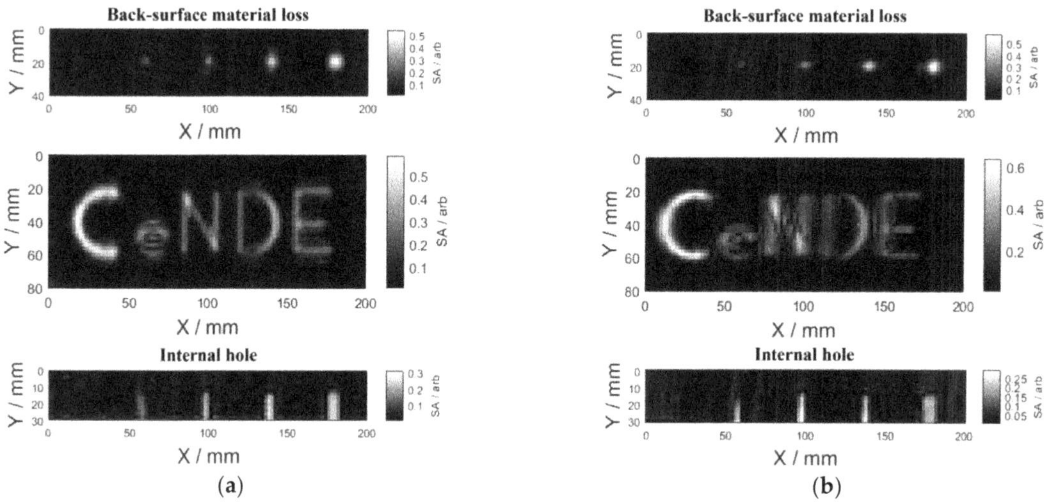

Figure 11. The reproduced defect images with the (**a**) parallel case and (**b**) orthogonal case.

Compared with Figures 8 and 11, it can be seen from Figure 11 that the contrast between the defect region and the flawless region in the defect image was markedly promoted. However, the smallest square back-surface material loss with a side length of 2 mm and the internal hole with a diameter of 1 mm are still not presented in Figure 11, and the letter 'e' for back-surface material loss was relatively blurry and inaccurate. This is due to the comprehensive effect of the transverse resolution of Ka-band microwave for GFRP materials detection (the half wavelength of 40 GHz microwave in GFRP: 1.8 mm), the minimum size of the Ka-band waveguide aperture (3.56 mm), and the scanning step size (2 mm), resulting in limiting the resolution of the defect image.

The contrast ratio between the two defect images in Figure 11 was calculated to quantitatively compare the influence of the microwave polarization direction with respect to the GFRP fiber direction on the testing sensitivity, which is calculated by the ratio of the mean values of pixels in the defect region to that of the flawless region. The results are listed in Table 4. It can be found from Table 4 that the contrast ratio of defect images was higher when the wave-polarization direction is parallel to the fiber direction in testing.

Table 4. Contrast ratio of defect images at the two wave-polarization angles.

Wave-Polarization Angle	Parallel Case	Orthogonal Case
Contrast ratio	5.7765	4.9937

3.4. Assessment of the Defect in-Plane Area

Based on the acquired defect images exhibited in Figure 11, a defect edge identification method is employed to obtain a quantitative evaluation of the in-plane area of buried defects. To obtain an accurate area value, it is critical to smooth the image by interpolation, and bicubic interpolation was used to increase the number of pixel points in the defect image. Subsequently, linear grayscale transformation was performed to enhance defect features in large-scale pixel images, followed by the Sobel operator being utilized to identify the defect edge in the defect image after linear grayscale transformation. The area enclosed by the defect edge curve is considered the in-plane area of buried defects. This process is abbreviated as the defect edge identification method.

The processing procedures include the following:

- Step 1: image interpolation. The number of pixel points is expanded from $W \times H$ to $(nW) \times (nH)$, so the image resolution unit is reduced from A to A/n^2. The number of pixel points in Figure 11 is increased 16 times, as shown in Figure 12a,d, along with the resolution unit reducing from 2 mm × 2 mm to 0.5 mm × 0.5 mm;
- Step 2: feature enhancement. The image obtained from step 1 is converted into a grayscale image with pixel values ranging from 0 to 255, and then linear transformation on the grayscale image is performed as Equation (10):

$$\mathbf{J} = (\mathbf{I} - \tau)/\gamma \tag{10}$$

  where, $\mathbf{J}$ is the pixel-value matrix of the defect image after linear grayscale transformation. $\mathbf{I}$ is the pixel-value matrix of the grayscale image obtained from step 1. If the value of a certain point in $\mathbf{J}$ is less than 0, it is considered as 0, similarly, if it is greater than 255, it is considered as 255. The coefficients, i.e., τ and γ are set as 64 and 11, respectively, in Equation (10) to enhance the defect features in Figure 12a,d. The enhancement results are shown in Figure 12b,e. It can be seen that the linear grayscale transformation method can make the small values in the defect image smaller and the large values larger, thereby further suppressing the background, enhancing the pixel values at the defect region, and thus the defect features are enhanced;
- Step 3: defect edge recognition. As the Sobel operator performs weighted processing on the influence of pixel positions, it can better handle images with grayscale gradients and high noise levels. The Sobel operator is utilized to identify the defect edges in Figure 12b,e, and the identification results are shown in Figure 12c,f. The figure shows that the final defect-edge recognition effect of back surface material loss of letter shape with the microwave polarization direction parallel to the fiber direction is better and clearer than that with the orthogonal scenario.

The in-plane areas of eight detected defects with known sizes, i.e., the defects #2~#5, #12~#15, were evaluated by using the proposed defect edge identification method. The evaluation results are shown in Figure 13.

It can be observed from Figure 13 that when the wave polarization direction is parallel to the GFRP fiber direction, the evaluated area value is closer to the actual area of the buried defect. By further calculation, the averaged relative error of evaluation results with the parallel scenario was 10.21% while the averaged relative error with the orthogonal scenario was 14.04%, which indicates that when the wave polarization direction is parallel to the GFRP fiber direction, the detection accuracy is higher. It is noteworthy that the pre-processing of the raw defect image by using the proposed direct wave suppression method is crucial and beneficial to the high evaluation accuracy regarding the in-plane area of the buried defect.

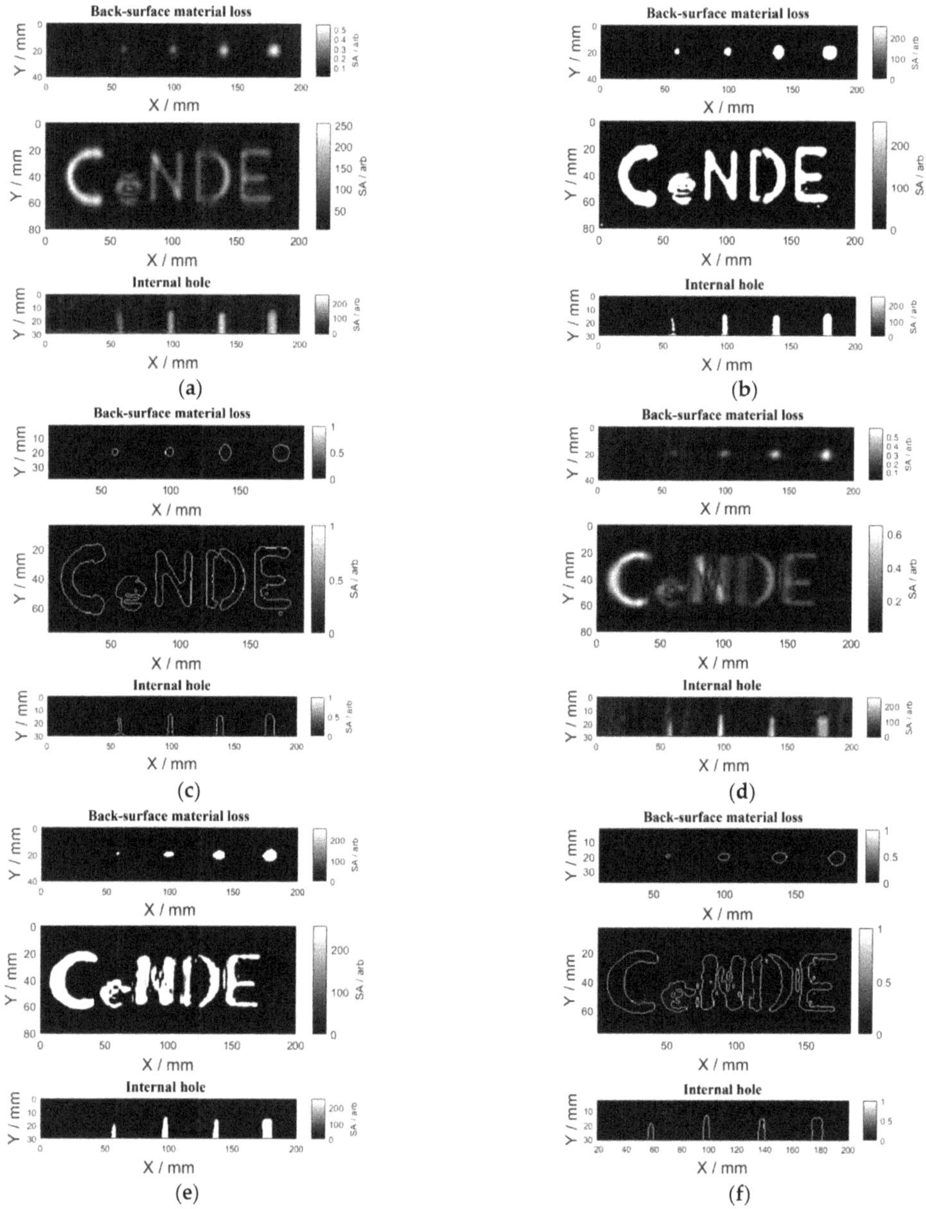

Figure 12. The proposed defect edge identification method: (**a**) image interpolation with the parallel case; (**b**) feature enhancement with the parallel case; (**c**) defect edge recognition with the parallel case; (**d**) image interpolation with the orthogonal case; (**e**) feature enhancement with the orthogonal case; (**f**) defect edge recognition with the orthogonal case.

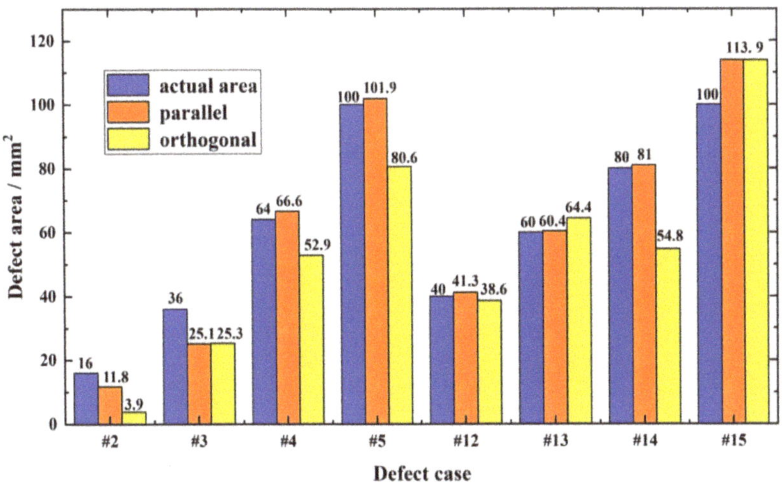

Figure 13. Assessment results of the in-plane areas of buried defects in GFRP.

4. Concluding Remarks

In this paper, the quantitative evaluation of buried defects in a unidirectional GFRP was intensively investigated via microwave reflectometry. The influence of two typical angles between the microwave polarization direction and the GFRP fiber direction, i.e., parallel and orthogonal on the testing sensitivity, was analyzed through simulations and experiments. Based on the equivalent relative permittivity corresponding to two angle scenarios calculated via theoretical analysis, a numerical simulation model was established to investigate the response of reflected signals to buried defects. In the experimental study, a microwave-reflectometry system was built for the imaging and quantitative evaluation of buried defects in GFRP. A direct wave suppression method and a defect edge identification method were proposed to improve the quality of the defect image and assess the in-plane area of buried defects, respectively. The investigation results indicated the following:

1. The simulated and experimental signals imply that regardless of the flawless region or the defect center, when the wave polarization direction is parallel to the fiber direction, the amplitude of S_{11} is greater than that with the orthogonal scenario;
2. The proposed direct wave suppression method can effectively suppress the direct-wave components in testing signals, significantly amplify the difference of signals between the defect center and the flawless region, and upgrade the quality of defect images. The minimum size of the detected defect can reach 2 mm;
3. The proposed defect edge identification method can availably evaluate the in-plane area of buried defects in GFRP. Both the contrast ratio of the defect image and the averaged relative error of assessment results regarding the defect in-plane area with two-angle scenarios imply that when the wave-polarization direction is parallel to the fiber direction, the testing sensitivity for GFRP buried defects is higher.

Author Contributions: Conceptualization, Y.F.; Data curation, Q.G. and X.Y.; Methodology, Y.L. and Y.F.; Investigation, R.W.; Formal analysis, R.W. and Q.G.; Writing—original draft, R.W.; Writing—review & editing, R.W., Y.F. and Y.L.; Supervision, Y.L. and Z.C.; Funding acquisition, Y.F., Y.L. and Z.C. All authors have read and agreed to the published version of the manuscript.

Funding: This work is supported in part by the National Magnetic Confinement Fusion Program of China under Grant 2019YFE03130003, National Natural Science Foundation of China under Grant 52107009, National Natural Science Foundation of China under Grant 11927801, Key R&D Project in Shaanxi Province under Grant 2023-YBGY-411.

Institutional Review Board Statement: Not applicable.

Informed Consent Statement: Not applicable.

Data Availability Statement: The data presented in this study are available on request from the corresponding author after obtaining permission of an authorized person.

Conflicts of Interest: The authors declare no conflict of interest.

References

1. Hollaway, L.C. A review of the present and future utilisation of FRP composites in the civil infrastructure with reference to their important in-service properties. *Constr. Build. Mater.* **2010**, *24*, 2419–2445. [CrossRef]
2. Wu, Z.; Xin, W.; Zhao, X.; Noori, M. State-of-the-art review of FRP composites for major construction with high performance and longevity. *Int. J. Sustain. Mater. Struct. Syst.* **2014**, *1*, 201–231. [CrossRef]
3. Wang, X.; Ding, S.; Qiu, A.; Ashour, A.; Wang, Y.; Han, B.; Ou, J. Improving bond of fiber-reinforced polymer bars with concrete through incorporating nanomaterials. *Compos. Part B Eng.* **2022**, *239*, 109960. [CrossRef]
4. Das, R.K.; Nayak, B.; Gautam, S.S.; Rana, N.K. Fabrication and characterisation of GFRP composite as skin material of sandwich structure. *Mater. Today Proc.* **2023**, *76*, 569–572. [CrossRef]
5. Aravind, N.; Samanta, A.K. Theoretical analysis on flexural strengthening of RC beams using GFRP composites in terms of strength to the weight ratio: A way forward. *Mater. Today Proc.* **2023**, *in press*. [CrossRef]
6. Reis, J.M.L.; Martins, F.D.F.; Mattos, H.S.D.C. Influence of ageing in the failure pressure of a GFRP pipe used in oil industry. *Eng. Fail. Anal.* **2017**, *71*, 120–130. [CrossRef]
7. Fang, H.; Bai, Y.; Liu, W.; Qi, Y.; Wang, J. Connections and structural applications of fibre reinforced polymer composites for civil infrastructure in aggressive environments. *Compos. Part B Eng.* **2019**, *164*, 129–143. [CrossRef]
8. Manalo, A.; Aravinthan, T.; Fam, A.; Benmokrane, B. State-of-the-art review on FRP sandwich systems for lightweight civil infrastructure. *J. Compos. Constr.* **2017**, *21*, 04016068. [CrossRef]
9. Seifoori, S.; Parrany, A.M.; Mirzarahmani, S. Impact damage detection in CFRP and GFRP curved composite laminates subjected to low-velocity impacts. *Compos. Struct.* **2021**, *261*, 113278. [CrossRef]
10. Castellano, A.; Fraddosio, A.; Piccioni, M.D. Quantitative analysis of QSI and LVI damage in GFRP unidirectional composite laminates by a new ultrasonic approach. *Compos. Part B Eng.* **2018**, *151*, 106–117. [CrossRef]
11. Asokkumar, A.; Jasiuniene, E.; Raisutis, R.; Kazys, R.J. Comparison of ultrasonic non-contact air-coupled techniques for characterization of impact-type defects in pultruded GFRP composites. *Materials* **2021**, *14*, 1058. [CrossRef] [PubMed]
12. Wu, Z.; Tao, N.; Zhang, C.; Sun, J. Prediction of defect depth in GFRP composite by square-heating thermography. *Infrared Phys. Technol.* **2023**, *130*, 104627. [CrossRef]
13. Garcea, S.C.; Wang, Y.; Withers, P.J. X-ray computed tomography of polymer composites. *Compos. Sci. Technol.* **2018**, *156*, 305–319. [CrossRef]
14. Li, Z.; Haigh, A.; Soutis, C.; Gibson, A.; Wang, P. A review of microwave testing of glass fibre-reinforced polymer composites. *Nondestruct. Test. Eval.* **2019**, *34*, 429–458. [CrossRef]
15. Li, Z.; Zhou, L.; Lei, H.; Pei, Y. Microwave near-field and far-field imaging of composite plate with hat stiffeners. *Compos. Part B Eng.* **2019**, *161*, 87–95. [CrossRef]
16. Mei, H.; Chen, J.; Fu, X.; Wang, L.; Guan, Z. Microwave reflection method for composite insulator internal defect detection. In Proceedings of the 12th IET International Conference on AC and DC Power Transmission (ACDC 2016), Beijing, China, 28–29 May 2016.
17. Wang, P.; Zhou, L.; Liu, G.; Pei, Y. In situ near-field microwave characterization and quantitative evaluation of phase change inclusion in honeycomb composites. *NDT E Int.* **2021**, *121*, 102469. [CrossRef]
18. Yu, Y.; Li, Y.; Qin, H.; Cheng, X. Microwave measurement and imaging for multiple corrosion cracks in planar metals. *Mater. Des.* **2020**, *196*, 109151. [CrossRef]
19. Rahman, M.S.U.; Haryono, A.; Abou-Khousa, M.A. Microwave non-destructive evaluation of glass reinforced epoxy and high density polyethylene pipes. *J. Nondestruct. Eval.* **2020**, *39*, 26. [CrossRef]
20. Shrifan, N.H.M.M.; Jawad, G.N.; Isa, N.A.M.; Akbar, M.F. Microwave nondestructive testing for defect detection in composites based on k-means clustering algorithm. *IEEE Access* **2021**, *9*, 4820–4828. [CrossRef]
21. Wang, P.; Li, Z.; Zhou, L.; Pei, Y. Microwave nondestructive detection and quantitative evaluation of kissing defects in GFRP laminates. *Compos. Sci. Technol.* **2018**, *162*, 117–122. [CrossRef]
22. Sutthaweekul, R.; Tian, G.Y.; Wang, Z.; Ciampa, F. Microwave open-ended waveguide for detection and characterisation of FBHs in coated GFRP pipes. *Compos. Struct.* **2019**, *225*, 111080. [CrossRef]
23. Shrifan, N.H.M.M.; Akbar, M.F.; Isa, N.A.M. Maximal overlap discrete wavelet-packet transform aided microwave nondestructive testing. *NDT E Int.* **2021**, *119*, 102414. [CrossRef]
24. Kharkovsky, S.; Zoughi, R. Microwave and millimeter wave nondestructive testing and evaluation—Overview and recent advances. *IEEE Intum. Meas. Mag.* **2007**, *10*, 26–38. [CrossRef]

25. Li, Z.; Haigh, A.; Soutis, C.; Gibson, A.; Sloan, R. Dielectric constant of a three-dimensional woven glass fibre composite: Analysis and measurement. *Compos. Struct.* **2017**, *180*, 853–861. [CrossRef]
26. Ida, N. The electromagnetic field equations and theoretical aspects. In *Microwave NDT*; Springer: Berlin/Heidelberg, Germany, 1992; pp. 1–38.
27. Mirotznik, M.S.; Yarlagadda, S.; McCauley, R.; Pa, P. Broadband electromagnetic modeling of woven fabric composites. *IEEE Trans. Microw. Theory Tech.* **2011**, *60*, 158–169. [CrossRef]
28. Liu, C.; Song, C.; Lu, Q. Random noise de-noising and direct wave eliminating based on SVD method for ground penetrating radar signals. *J. Appl. Geophys.* **2017**, *144*, 125–133. [CrossRef]

sensors

MDPI

Article

Visual Quantitative Detection of Delamination Defects in GFRP via Microwave

Xihan Yang, Yang Fang *, Ruonan Wang, Yong Li and Zhenmao Chen *

State Key Laboratory for Strength and Vibration of Mechanical Structures, Shaanxi Engineering Research Centre of NDT and Structural Integrity Evaluation, School of Aerospace Engineering, Xi'an Jiaotong University, Xi'an 710049, China; yxihan@stu.xjtu.edu.cn (X.Y.); w2502209459@stu.xjtu.edu.cn (R.W.); yong.li@mail.xjtu.edu.cn (Y.L.)

* Correspondence: fangyang@xjtu.edu.cn (Y.F.); chenzm@mail.xjtu.edu.cn (Z.C.)

Abstract: Glass Fiber reinforced polymers (GFRPs) are widely used and play an important role in modern society. The multilayer structure of GFRPs can lead to delamination defects during production and service, which can have a significant impact on the integrity and safety of the equipment. Therefore, it is important to monitor these delamination defects during equipment service in order to evaluate their effects on equipment performance and lifespan. Microwave imaging testing, with its high sensitivity and noncontact nature, shows promise as a potential method for detecting delamination defects in GFRPs. However, there is currently limited research on the quantitative characterization of defect images in this field. In order to achieve visual quantitative nondestructive testing (NDT), we propose a 2D-imaging visualization and quantitative characterization method for delamination defects in GFRP, and realize the combination of visual detection and quantitative detection. We built a microwave testing experimental system to verify the effectiveness of the proposed method. The results of the experiment indicate the effectiveness and innovation of the method, which can effectively detect all delamination defects of 0.5 mm thickness inside GFRP with high accuracy, the signal-to-background ratio (SBR) of 2D imaging can reach 4.41 dB, the quantitative error of position is within 0.5 mm, and the relative error of area is within 11%.

Keywords: nondestructive testing; microwave imaging; quantitative characterization; GFRP

1. Introduction

Glass fiber reinforced polymer (GFRP) is a lightweight, high specific strength, corrosion-resistant, electrothermally insulating, and transient ultra-high temperature resistant composite material that is widely used in critical structures across various engineering fields, such as aerospace, energy and power, and transportation [1–3]. However, due to its laminated structure, GFRPs are susceptible to interlaminar shear stress during processing, assembly, and service, which can result in internal delamination defects that compromise the structural integrity and safety of the material [4,5]. These defects can include delamination between layers and separation of panels, which can lead to a range of potential consequences, such as reduced load-bearing capacity, increased risk of catastrophic failure, and compromised service life. Therefore, it is essential to conduct nondestructive testing and quantitative characterization of GFRP internal delamination defects in order to ensure engineering safety.

Nondestructive testing (NDT) methods for GFRPs currently include ultrasonic testing [6,7], infrared imaging testing [8], radiographic testing [9], and terahertz testing [10–12]. While each method has its own advantages in actual detection, there are also limitations to consider. Ultrasonic detection cannot achieve noncontact detection, and the choice of coupling agent can greatly impact the detection signal. Radiographic detection equipment is expensive and may not provide enough penetration depth for thicker test pieces. Infrared imaging methods are susceptible to signal interference from surrounding environment radiation. Terahertz detection is limited by its penetration depth, making it challenging to detect thicker workpieces.

Microwave imaging nondestructive testing is a technique that utilizes the difference in microwave response of defects and dielectric materials. It achieves imaging reconstruction and defect characterization through signal processing and parameter extraction from the reflected signal of the target. It is widely used in nonmetal detection and multilayer structure detection due to its high penetration capability, high resolution, large detection depth, noncontact nature, and high efficiency [13,14]. Currently, there are two primary methods for detecting internal defects in nonmetallic materials. The first method is microwave imaging detection, which utilizes imaging algorithms such as wave number domain imaging [15,16], holographic imaging [17], time domain reflectometry [18], and other methods such as unsupervised machine learning methods [19,20]. Microwave imaging detection methods enable visual detection of defects. The second method is microwave coefficient detection, which involves processing the amplitude and phase of the echo signal [21–24]. By analyzing these characteristics, the location, size, and other parameters of defects can be derived. Microwave coefficient detection methods allow for quantitative assessment of defects. While current microwave nondestructive testing research has achieved significant results in detection and imaging of microscopic defects, there are still gaps in the following areas of research. Firstly, in signal processing algorithms, the current imaging preprocessing filtering and clutter suppression processing is insufficient, leading to poor contrast and indistinguishable defect characteristics. Secondly, in imaging algorithms, the traditional method of feature extraction primarily focuses on imaging, resulting in suboptimal utilization of electromagnetic wave information and low imaging quality. Moreover, the current studies predominantly focus solely on detecting internal defects in GFRP and do not have automated quantitative characterization of defects. To address this issue and effectively detect and accurately characterize internal delamination defects in GFRP, we propose a method based on spectral reconstruction that allows for fast imaging and precise quantitative characterization.

Firstly, in order to suppress strong interference signals and enhance target features, we propose a clutter suppression method based on adaptive singular value decomposition. This method addresses the problem of inaccurate signal subspace division by traditional SVD methods and improves the signal-to-noise ratio (SBR) by adjusting the singular value sequence according to the image standard deviation. Images with good contrast and high quality serve as the foundation for achieving accurate quantitative assessment of defects. Secondly, to obtain a 2D microwave image of the target, we developed a wave spectrum reconstruction algorithm that enables visual detection of the target. The algorithm is based on the 2D Fourier transform and reconstructs the signal distribution at the target plane by performing phase compensation on the signal in the wave number domain. Thirdly, in order to quantitatively characterize defects, we developed the OTSU image segmentation algorithm to accurately separate the target from the background. This algorithm utilizes grayscale traversal to search for the optimal threshold value that maximizes the interclass variance between the target and background regions. By applying the Regionprops function, the pixel value distribution of the defect area is calculated, allowing us to obtain the location and area information of the defect and achieve precise characterization of the internal layered defects of GFRP. Finally, we built a microwave testing experimental system and scanned the designed specimens to verify the effectiveness of the proposed method and analyze its performance.

2. Methods of Spectral Reconstruction Imaging and Quantitative Characterization

As illustrated in Figure 1, there is continuous movement of the control probe along the x' and y' directions, transmitting and receiving microwave signals within the scanning range $L_x \times L_y$. The GFRP specimen contains delamination defects and is located at a distance R_0 from the scanning plane where the control probe is positioned. During the 2D scanning process, the scattering at each sampling position is recorded as a matrix, with each element of the matrix representing the complex echo signal $s(x', y')$ of a specific sampling coordinate point.

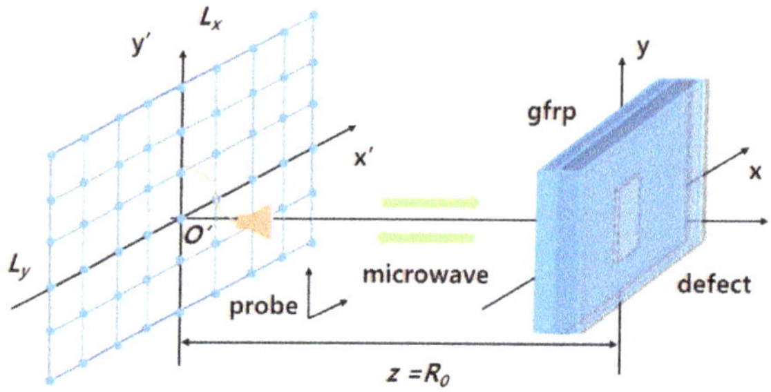

Figure 1. Schematic diagram of 2D scanning microwave detection.

In the actual SISO (single-input single-output) testing, the echo signal $s(x', y')$, i.e., S11 parameter collected by a vector network analyzer (VNA), represented by Equation (1), is related to various parameters such as the signal frequency f, the specimen thickness d, and the reflection coefficient Γ at the heterogeneous interface. The value of Γ can be obtained from the wave impedance Z_S, where Z_s is a function of magnetic permeability μ and dielectric constant ε. As ε of air and GFRP materials are distinct, when the microwave signal propagates to the interface between air and GFRP, it will be reflected [25]. Therefore, when there are delamination defects inside GFRP, the echo signal $s(x', y')$ will be changed.

$$S11 = \frac{\Gamma(1 - e^{2jkd})}{1 - \Gamma^2 e^{2jkd}} \qquad \Gamma = \frac{Z_S - 1}{Z_S + 1} \qquad Z_S = \sqrt{\frac{\mu}{\varepsilon}} \tag{1}$$

As illustrated in Figure 2, when the probe irradiates the GFRP specimen, the signal propagates through the air layer and reaches the surface of the GFRP specimen, and a portion of the signal is reflected back, while another portion continues to propagate through the air–GFRP interface and into the GFRP specimen. As the signal encounters internal delamination defects, it is reflected once again, and the received signal is a combination of coupled and reflected signals, which is then picked up by the probe and recorded by the system as $s(x', y')$.

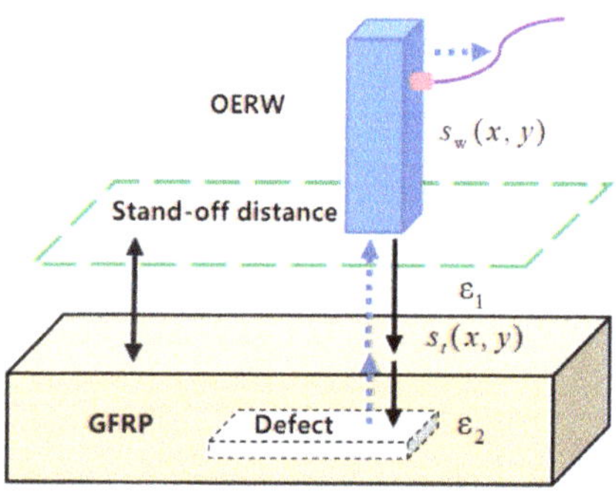

Figure 2. Principle of microwave signal reflection.

$s(x', y')$ consists of two main components:

$$s(x', y') = s_w(x', y') + s_t(x', y') \tag{2}$$

The signal $s_w(x', y')$ represents the interference signal, which includes the echo signal at the probe antenna coupling, as well as any other unwanted signals such as direct wave.

The probe-coupled reflected wave is the signal caused by the impedance mismatch at the connection when the microwave propagates from VNA to the probe through a cable. Direct wave refers to the microwave signal reflected on the surface of the GFRP specimen when the microwave is emitted from the probe through the air. $s_t(x', y')$ represents the target signal. In order to achieve 2D imaging of the internal delamination defects of GFRP, the target signal is therefore analyzed. As depicted in Figure 1, the coordinates of the probe are represented as (x', y'), while the target image function of the measured target is denoted by $g(x, y)$. Then the echo signal at (x', y') is

$$s_t(x', y') = g(x, y) \cdot \exp(-2j\pi \cdot 2Rf/c) \tag{3}$$

$$R = \sqrt{(x' - x)^2 + (y' - y)^2 + {R_0}^2} \tag{4}$$

The primary objective of 2D imaging is to effectively reconstruct the 2D scattering distribution and determine the target image function $g(x, y)$ [26]. To achieve quantitative characterization of defects, the function $g(x, y)$ is first normalized to function $g'(x, y)$. Then, the target region A of the 2D image $g'(x, y)$ is separated using threshold segmentation. Finally, the image function $g'(x, y)[(x, y) \in A]$ is binarized to function $g(x, y)[(x, y) \in A]$ to enable visual detection of defects. To achieve visual detection and quantitative characterization of internal delamination defects in GFRP, a three-part algorithm based on spectral reconstruction was developed. The algorithm consists of clutter suppression, 2D imaging, and quantitative characterization.

2.1. Adaptive Singular Value Decomposition-Based Clutter Suppression

In the processing of the echo signal, achieving high-quality 2D imaging of the target requires initial clutter suppression to improve the signal-to-noise ratio (SNR). Common clutter suppression methods include coherent processing filtering, time-domain filtering, and wavelet denoising. The approach of clutter suppression based on singular value decomposition (SVD) is a subspace algorithm that involves decomposing the signal vector space into two subspaces, where one is predominantly dominated by the target signal and the other by the noise signal. Clutter suppression is achieved by eliminating the signal vector components belonging to the noise-dominated subspace [27]. Let the echo signal $s(x', y')$ at each sampling point form an $m \times n$ dimensional matrix S of rank r. SVD of S is expressed as follows:

$$S = U \begin{bmatrix} \Sigma & 0 \\ 0 & 0 \end{bmatrix} V^H \tag{5}$$

where U and V are $m \times m$, $n \times n$ orthogonal matrices, respectively (m, n sampling points). $\sum$ is an $r \times r$ dimensional diagonal array whose diagonal elements are the nonzero singular values σ_i of matrix S and are arranged in nonincreasing order, i.e., $\sigma_1 \geq \sigma_2 \geq \cdots \geq \sigma_r$. Since the rank of matrix S is r, the streamlined form of the singular value decomposition of S can be obtained by removing the zero singular values from Equation (5).

$$S = \sum_{i=1}^{r} \sigma_i' \mu_i {v_i}^H \tag{6}$$

In Equation (6), the left and right eigenvectors of the matrix are represented by μ_i and v_i, respectively, and $\sum_i \sigma_i \mu_i {v_i}^H$ denotes the i subspace echo signal. By rewriting Equation (6) after vector subspace decomposition, we obtain

$$S = \sum_{i \in T} \sigma_i \mu_i {v_i}^H + \sum_{i \in N} \sigma_i \mu_i {v_i}^H \tag{7}$$

In Equation (7), T represents the set of target signal subspace, while N represents the set of clutter signal subspace. In the SVD-based clutter suppression technique, once the target and clutter subspaces have been divided, the corresponding singular values can be used to identify their respective subspace signal vectors. By setting the singular values corresponding to the clutter dominant subspace to zero, the corresponding clutter signal of subspace N can be effectively removed, thereby achieving the goal of clutter suppression. However, in practical applications of SVD filtering, accurate division of the signal subspace and clutter subspace can be a challenge, leading to inadequate removal of the clutter subspace vector. To address the issue of inaccurate signal subspace division in SVD-based clutter suppression, an improved method is proposed in this study. The new method is based on adaptive SVD, which involves processing the specific subspace after decomposition instead of simply removing the maximum singular value to suppress the clutter. By optimally adjusting the size of the corresponding singular value in the subspace, the SNR can be improved, leading to effective suppression of strong clutter.

Rewriting Equation (7) as Equation (8), where W_1 and W_2 denote the set of clutter in the first and second signal subspaces, T_1 and T_2 denote the set of target signals in the first and second signal subspaces. Both the direct wave and the probe-coupled reflected wave energies are much stronger than the target signal energy, and during the SVD process, the subspace is divided based on the energy ratio [28]. Thus, the strong clutter signal mainly exists in the subspace corresponding to the first singular value, while the target signal mainly exists in the subspace corresponding to the second singular value. Therefore, the first subspace echo signal contains a higher proportion of the interference signal, denoted as $W_1 > T_1$, while the second subspace echo signal contains a higher proportion of the target signal, denoted as $T_2 > W_2$. This approach enables the reduction of the first singular value by a factor of k_1 to suppress strong clutter and the amplification of the second singular value by a factor of k_2 to enhance the SNR of the target signal.

$$S = (\sigma_1/k_1)\mu_1 v_1{}^H + (\sigma_2/k_2)\mu_2 v_2{}^H + \sum_{i=3}^{r} \sigma_i \mu_i v_i{}^H \quad (8)$$

The selection of the optimization coefficients k_1 and k_2 for the singular values is determined based on the image *std* (standard deviation), which ranges from 1 to Q_1/Q_2 (Q_1 is the first singular value, Q_2 is the second singular value). The *std* reflects the degree of dispersion of the grayscale values of the image pixels around the mean value. If the image standard deviation is larger, it indicates that the grayscale values of the image pixels are more widely dispersed around the mean value, implying a better image quality. This means that the degree of separation between defects and background in the image is greater, resulting in a better suppression effect on the interference signal and a more optimal selection of the singular value optimization coefficients k_1 and k_2. The optimized experimental data, namely, the experimental data after clutter suppression, are subjected to 2D imaging. The standard deviation of the image is a function of the optimization coefficients k_1 and k_2. When the standard deviation of the image is the maximum, the values of k_1 and k_2 are selected most appropriately.

The *std* of image is calculated using the following formula:

$$std = \sqrt{\frac{1}{MN}\sum_{i=1}^{M}\sum_{j=1}^{N}(F(i,j) - u)^2} \quad (9)$$

In Equation (9), M and N represent the size of the image, $F(i,j)$ represents the pixel value at row i and column j, and u represents the mean value, which is calculated by summing all pixel values and dividing by the total number of pixels in the image. When the maximum value of image *std* is obtained, the corresponding singular value optimization coefficients k_1 and k_2 are determined, and the new singular value matrix is reconstructed with the corresponding singular vectors to derive the signal $s_l(x', y')$ in the corresponding spatial coordinates after clutter suppression. Figure 3 illustrates the flowchart of the adaptive SVD method.

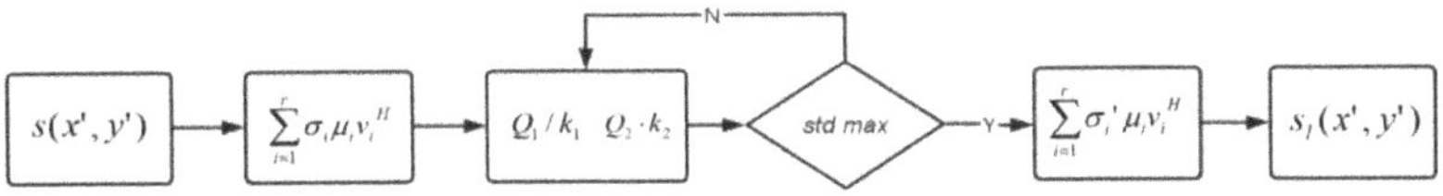

Figure 3. Clutter suppression method based on adaptive SVD.

2.2. 2D Imaging Method Based on Spectral Reconstruction

To achieve highly accurate detection results, it is crucial to obtain a high-resolution, 2D image with clear target features. Currently, numerous scholars have devoted significant efforts to achieve high-quality 2D imaging by enhancing techniques, such as backprojection algorithms [29], frequency scaling algorithms [30], and holographic imaging algorithms [31]. The objective is to improve the signal-to-noise ratio and enhance image features. Among the various microwave imaging algorithms available, the spectral reconstruction algorithm stands out because it does not require any approximation conditions in the inverse reconstruction process [32,33]. By accurately focusing on the entire imaging area based on the scattering point model, the spectral reconstruction algorithm is designed to enable 2D imaging of defects inside GFRP.

The fundamental concept underlying the spectral reconstruction imaging algorithm is derived from Equation (3). Specifically, this process can be modeled as a plane wave passing through a linear, spatially-invariant system, which acts as a linear, dispersive spatial filter with a finite spatial bandwidth. In this filter, the amplitude of the wave remains constant within the spatial frequency domain passband, while a phase shift is introduced. Therefore, the fundamental principle underlying the spectral reconstruction imaging algorithm involves applying phase compensation to microwave signals in the spatial wave number domain [34,35]. This approach enables the algorithm to solve for the target image function $g(x, y)$ at the location of the defect. Specifically, the filtered sampling data $s_l(x', y')$ can be reconstructed using the spectral reconstruction imaging algorithm to generate the image function $g(x, y)$, thereby enabling the identification of the defect.

The algorithm involves several main steps. Firstly, the received reflected waves are converted from the spatial domain to the wave number domain using a 2D Fourier transform. Next, a matched filter is constructed to filter the spatial wave spectrum. Finally, the scattering characteristic distribution of the target is obtained by Fourier inversion, allowing for the reconstruction of the scattering intensity distribution at the target location.

Firstly, the sampled data after clutter suppression are transformed into the spatial wave number domain using the 2D Fourier transform, and the microwave spatial spectrum $S(k_x, k_y, z = R_0)$ is derived:

$$S(k_x, k_y, z = R_0) = \int_{-L_x/2}^{L_x/2} \int_{-L_y/2}^{L_y/2} s_l(x', y') \exp[-j(k_x x + k_y y)] dx' dy' \tag{10}$$

In Equation (10), L_x and L_y correspond to the scanning range of the scanning plane, while k_x and k_y represent the spatial frequencies in the x and y directions, respectively.

Next, the spatial wave spectrum $S(k_x, k_y, z = R_0)$ undergoes phase compensation, after which the spectrum $S(k_x, k_y, z = 0)$ in the plane $z = 0$ of the defect can be obtained:

$$S(k_x, k_y; z = 0) = S(k_x, k_y; z = \mathrm{R}_0) \times H(k_x, k_y) \tag{11}$$

In Equation (11), $H(k_x, k_y)$ is the transfer function of the linear dispersive spatial filter used for phase compensation. It compensates for the phase by equating the plane wave components propagating in different directions from the position $z = z_0$ to the position $z = 0$. Its mathematical expression is as follows:

$$H(k_x, k_y) = circ(\frac{{k_x}^2 + {k_y}^2}{k^2}) \exp(jk_z R_0) \tag{12}$$

$$k_z = \sqrt{k^2 - k_x{}^2 - k_y{}^2} \tag{13}$$

$$circ(\frac{k_x{}^2 + k_y{}^2}{k^2}) = \begin{cases} 1 & k_x{}^2 + k_y{}^2 < k^2 \\ 0.5 & k_x{}^2 + k_y{}^2 = k^2 \\ 0 & k_x{}^2 + k_y{}^2 > k^2 \end{cases} \tag{14}$$

When calculating the transfer function, it is important to consider the distance between the defect in GFRP and the probe, denoted as $R_0 + h$, as well as the distance between the defect and the surface of the GFRP workpiece, denoted as h. The transfer function is processed in the frequency domain, as shown in Equations (15)–(17), where the spatial wave number is $k = 4\pi f/c$, f is the emitted signal frequency, c is the electromagnetic wave propagation velocity in the medium, and k_1 and k_2 are related to the air and GFRP dielectric constants, respectively.

$$H_1(k_x, k_y) = circ(\frac{k_x{}^2 + k_y{}^2}{k_1{}^2}) \exp(j\sqrt{k_1{}^2 - k_x{}^2 - k_y{}^2 R_0}) \tag{15}$$

$$H_1(k_x, k_y) = circ(\frac{k_x{}^2 + k_y{}^2}{k_2{}^2}) \exp(j\sqrt{k_2{}^2 - k_x{}^2 - k_y{}^2 h}) \tag{16}$$

$$H(k_x, k_y) = H_1(k_x, k_y) \cdot H_2(k_x, k_y) \tag{17}$$

Finally, the target scattering intensity at the corresponding spatial coordinate position, i.e., the target image function $g(x, y)$, is obtained by performing the 2D Fourier inverse transform of the spectrum $S(k_x, k_y, z = 0)$ at the target plane.

$$g(x, y) = \frac{1}{4\pi^2} \int\limits_{K_x} \int\limits_{K_y} S(k_x, k_y, z = 0) \exp(jk_x x + jk_y y) dk_x dk_y \tag{18}$$

The values of K_x, K_y in Equation (18) can be determined by using the relationship between the independent variables of the Fourier transform pair:

$$K_x = \pi/\Delta x' \qquad K_y = \pi/\Delta y' \tag{19}$$

The variables $\Delta x'$ and $\Delta y'$ in Equation (19) represent the step along the two coordinate axes. The flow of the spectral reconstruction algorithm is depicted in Figure 4.

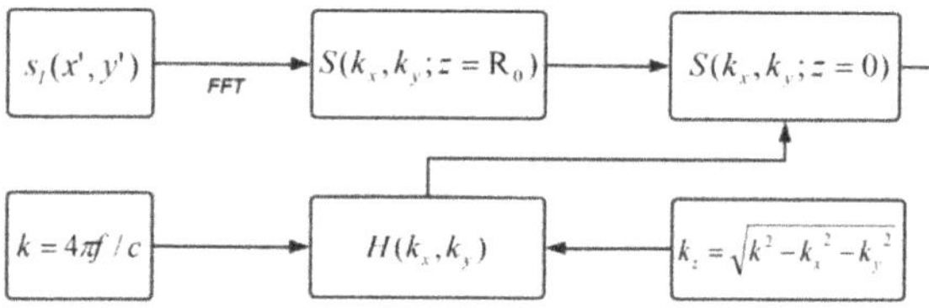

Figure 4. Spectral reconstruction algorithm.

2.3. *Quantitative Characterization Methods*

To enable the quantitative evaluation of defects, the image function $g(x, y)$ must first undergo processing to extract the target defect region A from the background region B of the 2D image G. Next, statistical analysis is performed on image regions$g(x, y)[(x, y) \in A]$ containing only defects, allowing for the characterization of defect locations in terms of their region center coordinates, as well as the quantification of defect size in terms of region area.

To perform quantitative evaluation of defects, several steps are taken. First, the pixel values of the image function $g(x,y)$ are normalized to obtain the function $g'(x,y)$. Second, the optimal threshold T for image segmentation is obtained using the OTSU algorithm [31]. Next, $g'(x,y)[(x,y) \in A]$ is binarized using this threshold to obtain the function $g(x,y)[(x,y) \in A]$, which contains only the defective regions A. Finally, the center-of-mass coordinates (x_i, y_i) of each part of the defective regions are obtained using the Regionprops function, and the number of pixels in each defective region is determined based on the center-of-mass coordinates. This pixel count is then converted into the defective area S_i to quantitatively characterize the defect location and area.

The OTSU algorithm, also known as the maximum between-class variance method, is based on the least squares principle and uses the grayscale histogram to achieve the best segmentation in a statistical sense. Its basic principle is to segment the grayscale value of the image into two parts using the best threshold so that the variance between the two parts is maximum, resulting in maximum separability. Let t be the segmentation threshold between the defect and the background. The number of defective pixel points as a proportion of the image is $w_0(A/G)$, with an average grayscale of u_0. Similarly, the number of background points as a proportion of the image is $w_1(B/G)$, with an average grayscale of u_1. The total average grayscale of the image is u, and v is the variance of the target defect and the background image:

$$u = w_0 \times u_0 + w_1 \times u_1 \tag{20}$$

$$v = w_0 \times (u_0 - u)^2 + w_1 \times (u_1 - u)^2 \tag{21}$$

Combining Equations (20) and (21):

$$v = w_0 \times w_1 \times (u_0 - u_1)^2 \tag{22}$$

The primary objective of the algorithm is to identify the optimal threshold T among the threshold t variables, which maximizes the interclass variance v between region A and region B. To improve the efficiency of the algorithm, a two-step approach of integer traversal and fractional traversal is employed to find the optimal threshold value through stepwise approximation. Firstly, the pixel values $g'(x,y)$ of the entire image region G are normalized to 255 levels of gray values. Then, the algorithm searches for the best integer threshold value T_N within the range of 1 to 255 integers. Subsequently, the search continues in the range of $T_N - 1$ to $T_N + 1$. At this point, the difference between the defective region A and the background region B is maximum, and T is the best threshold value. When the variance v reaches its maximum, it can be inferred that the difference between the defective region and the background region is also at its maximum. Hence, the threshold value T obtained at this point is considered to be the optimal threshold. Finally, the threshold T is used to segment the image function $g'(x,y)$ into $g'(x,y)[(x,y) \in A]$, which are then binarized such that the pixel values in the defective region A are set to 1 and the pixel values in the background region B are set to 0. This process produces a 2D image function $g(x,y)[(x,y) \in A]$ containing only the defects.

To further achieve a quantitative characterization of the location and area of defects, the pixel statistics function of Regionprops is applied to $g(x,y)[(x,y) \in A]$ in order to obtain the center-of-mass coordinates (x_i, y_i) and area S_i of each region, thus enabling a quantitative characterization of the defects. Figure 5 illustrates the flow of the quantitative characterization method.

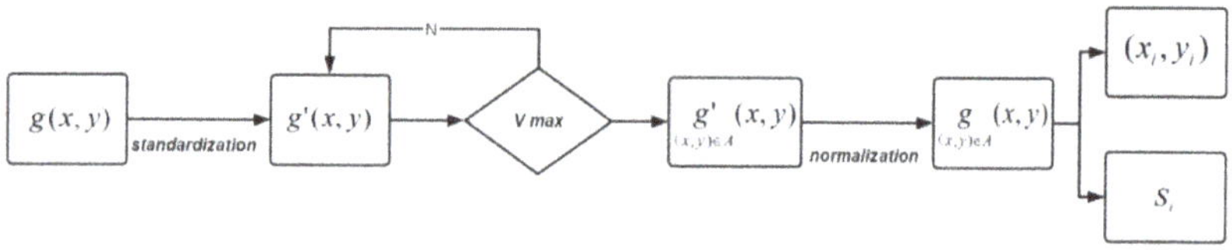

Figure 5. Quantitative characterization method.

2.4. Complete Process and Step-by-Step Description

The entire process can be divided into three steps: adaptive SVD clutter suppression processing, spectral reconstruction 2D imaging, and quantitative characterization based on the OTSU image segmentation algorithm. Figure 6 illustrates the overall flow chart of the process.

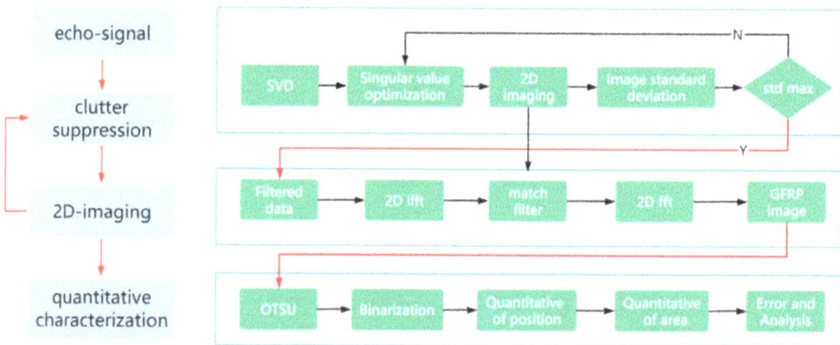

Figure 6. Complete process and step-by-step description.

In the clutter suppression process, we first perform SVD decomposition on the experimental data to obtain the singular value matrix. Next, we optimize the first two singular values based on the image standard deviation. When the standard deviation reaches its maximum, we solve for the new singular value matrix using the corresponding optimization coefficients. Finally, we multiply and reorganize the singular value matrix with the left and right singular vectors to complete the adaptive SVD clutter suppression processing.

In 2D imaging, the data undergo a 2D Fourier inversion to transform the spatial domain into the wave number domain. A linear dispersive spatial filter is then constructed on the wave number domain to match the filtered signal. Finally, the matched filtered data are transformed back to the spatial domain by the 2D Fourier transform to complete the wave-spectrum reconstruction for 2D imaging.

In quantitative characterization, the image is segmented using the OTSU image segmentation algorithm to find the segmentation threshold of defects and background. The image is then binarized based on this threshold to obtain a two-dimensional image containing only defects. The Regionprops function is used to find the center-of-mass coordinates of each defect, and the defect area is calculated based on these coordinates to complete the quantitative characterization of defect location and area.

3. Experimental Verification and Analysis

3.1. System Construction and Data Acquisition

As illustrated in Figure 7, we developed a 2D scanning microwave detection experimental system. This system comprises a vector network analyzer (VNA N5224A, from Agilent, Palo Alto, CA, USA), a Ka-band waveguide probe, a 3D automatic scanning stage, and automatic scanning control software. The VNA and waveguide probe are utilized to transmit and receive microwaves directionally in the microwave detection system. Additionally, the waveguide probe is moved by the 3D automatic scanning table to ensure that the microwave signal covers the surface of the specimen. The parameters of the automatic scanning control software are adjusted using a computer, and the experimental data are processed accordingly.

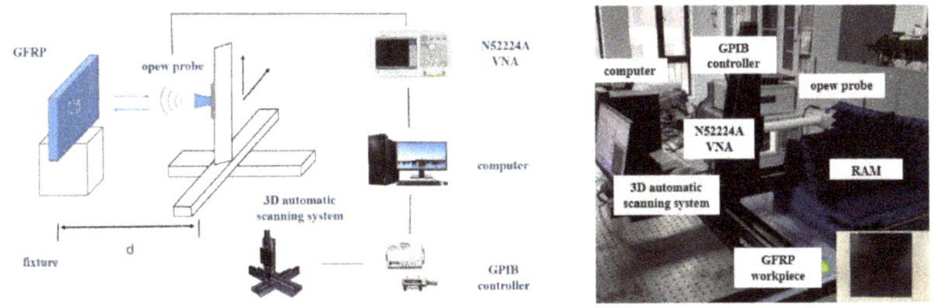

Figure 7. Microwave detection system.

The experiment is conducted on a GFRP specimen with internal delamination defects. A 0.5 mm thick PTFE film is used to simulate the delamination defects, and Figure 8 presents information regarding the location and area of the defects.

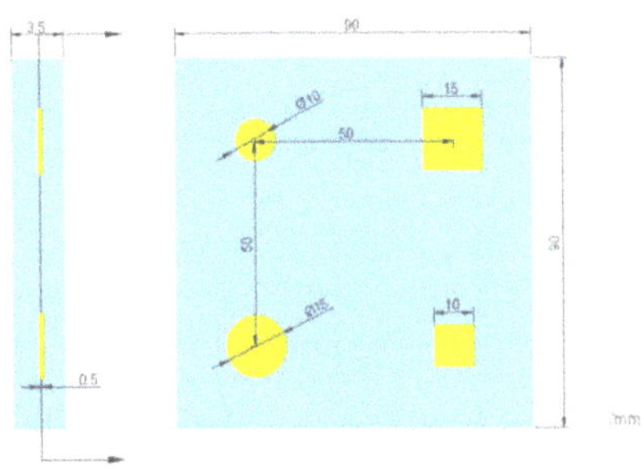

Figure 8. Design of GFRP specimen.

To begin the sweeping experiment, it is necessary to determine the appropriate emission band. In order to achieve a balance between microwave penetration and resolution in the 2D scanning, we selected the Ka band (26.5–40 GHz) for our experiment. The distance between the waveguide probe and the surface of the specimen was set to 10 mm. If the lift-off is too small, the microwave signal cannot approximate a plane wave, which can affect the wave spectrum reconstruction into an image. Conversely, if the lift-off is too large, the waveguide probe receives less reflected signal. The automatic scanning range was determined based on the size of the workpiece and the location of the buried defects, and was set at 90 mm × 90 mm. In order to ensure efficient experimental scanning and accurate image reconstruction, it is crucial to determine a suitable scanning step. Typically, the scanning step should not be less than a quarter of the wavelength, which is calculated at the mid-frequency point of the Ka band, to avoid a decrease in scanning efficiency or a lack of data for image reconstruction. Lastly, the number of scan points was calculated based on the scan range and step distance, and set at 46 × 46. The experimental parameters are shown in Table 1.

Table 1. Experimental parameters.

Microwave Frequency Band	Stand-Off Distance	Scanning Range	Scanning Step	Scanning Points	Frequency Point of 2D Imaging
Ka (26.5–40 GHz)	10 mm	90 × 90 mm	2 mm	46 × 46	39.5 GHz

3.2. Data Processing and Imaging

After collecting the experimental data, 2D imaging was performed using the spectral reconstruction algorithm without any clutter suppression processing. The resulting image, as shown in Figure 9a, exhibits streaks that interfere with the accurate quantitative characterization of the target defects. When the traditional SVD method is used for filtering, the imaging results are shown in Figure 9b. The defect features are not obvious and the contrast is poor. To address this issue, the echo signal was subjected to an adaptive SVD clutter suppression process to reduce the interference from clutter signals prior to 2D imaging. The developed algorithm is imaged under near-field conditions. However, it can theoretically also be applied to far-field imaging, but current research focuses on near-field imaging due to the significant impact of stand-off, resulting in relatively poor far-field results.

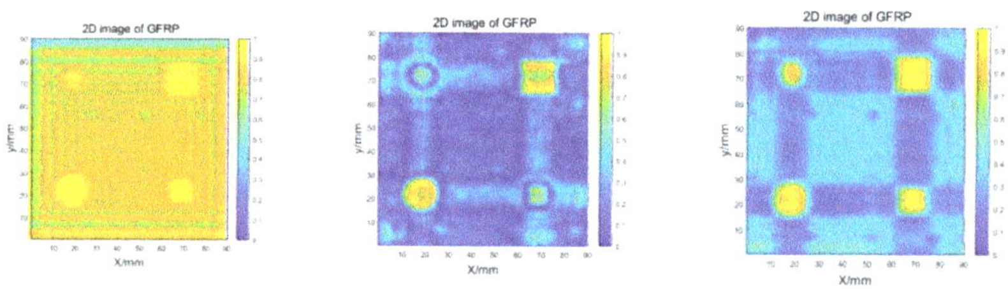

(a) (b) (c)

Figure 9. Comparison of imaging results: (**a**) without filtering; (**b**) traditional SVD method; (**c**) adaptive SVD method.

In the adaptive SVD clutter suppression process, the first two singular values are optimized based on the method described in Section 2.1, and the corresponding *std* of the images with different optimization coefficients are computed. Figure 10c displays the results of the singular value optimization, with a maximum *std* of 0.167, corresponding to the first singular value optimization coefficient k_1 of 7.2 and the second singular value optimization coefficient k_2 of 2.2. The optimized singular value matrix and singular vector are multiplied and recombined to obtain the clutter-suppressed echo signal, which is then used for 2D imaging using the spectral reconstruction algorithm. The imaging results are shown in Figure 9c. Compared to other imaging methods or clutter suppression techniques, our proposed combination of adaptive SVD and spectral reconstruction, along with techniques such as PCA [15] and SAR [16] imaging, offers superior imaging quality and contrast. These enhancements enable clearer visualization and differentiation of defects from the background, allowing for more precise measurements and evaluations.

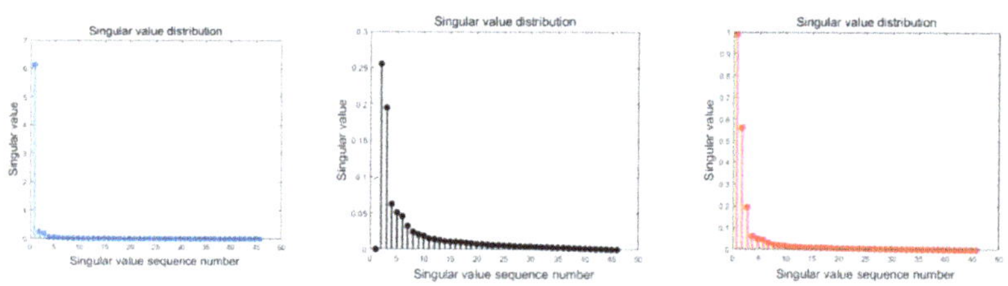

(a) (b) (c)

Figure 10. Singular value: (**a**) without filtering; (**b**) traditional SVD method; (**c**) adaptive SVD method.

3.3. Quantitative Characterization and Analysis of Results

As illustrated in Figure 11, we conducted grayscale histogram statistics of the 2D image that was formed. Based on this analysis and the OTSU image segmentation algorithm, we determined the segmentation threshold T to be 0.561. To further evaluate the quality of the image, we introduced signal-to-background ratio (SBR) of image. This concept was derived from spectroscopy and measures the level of significance of the target signal. It provides an intuitive reflection of the ratio between the grayscale value of the target signal and that of the background [32]. Equation (23) is used to calculate this ratio, where $f_s(x,y)$ denotes the mean gray value of the signal region, and $f_b(x,y)$ denotes the mean gray value of the background region. In situations where the gray value of the signal is equal to that of the background, this indicates that the signal and background cannot be differentiated. As a result, a larger SBR indicates better image quality. Using the determined threshold value T, we can solve for $SBR = 4.41\text{dB}$.

$$SBR = 10\left|\lg\frac{f_s(x,y)}{f_b(s,y)}\right| \tag{23}$$

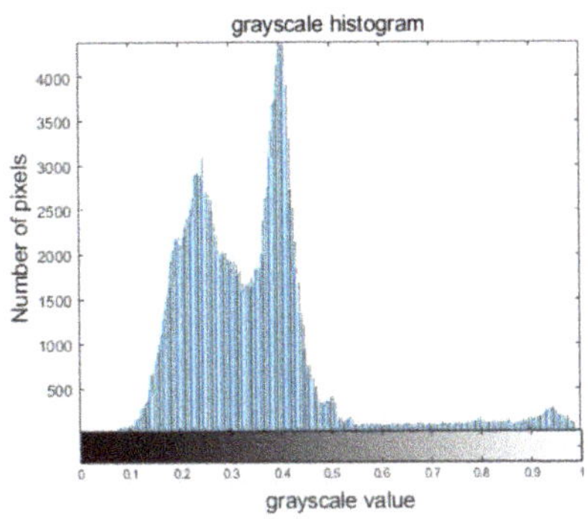

Figure 11. Grayscale histogram.

Based on the previously determined threshold value T, we performed further binarization of the GFRP image to eliminate the background information and obtain a 2D image that displays only the defects. The resulting image of delamination defects is presented in Figure 12, and it can be observed that the shapes of the defects depicted in the image are in good agreement with the anticipated design defect shapes.

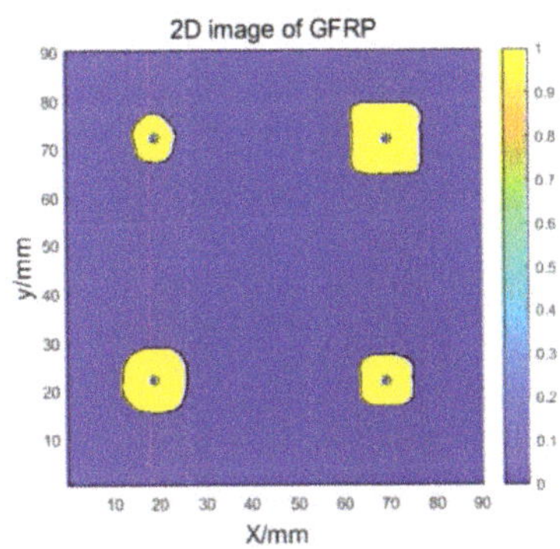

Figure 12. Delamination defects.

In order to characterize the number and location of defects, we conducted pixel statistics using the Regionprops function to extract the center-of-mass coordinates (x_i, y_i) for each defect present in the image. The resulting center-of-mass coordinates were then utilized to determine the relative location information of the defects, and the obtained results are tabulated in Table 2.

Table 2. Position detection results.

	Position 1	Position 2	Position 3	Position 4
Theoretical (mm)	(0, 0)	(0, 50)	(50, 50)	(50, 0)
Detection (mm)	(0, 0)	(0.17, 49.93)	(50.41, 49.80)	(50.27, −0.28)
Error (mm)	(0, 0)	(0.17, 0.07)	(0.41, 0.20)	(0.27, 0.28)

To further characterize the defect area information, we determined the number of pixels for each individual defect part based on its center-of-mass coordinates. Given that the side length of each pixel block is 2 mm (scanning step), we were able to convert the number of pixels into an area measurement S_i. We compared the characterized area with the design area to evaluate the accuracy of our defect area characterization method. The resulting evaluation errors are presented in Figure 13.

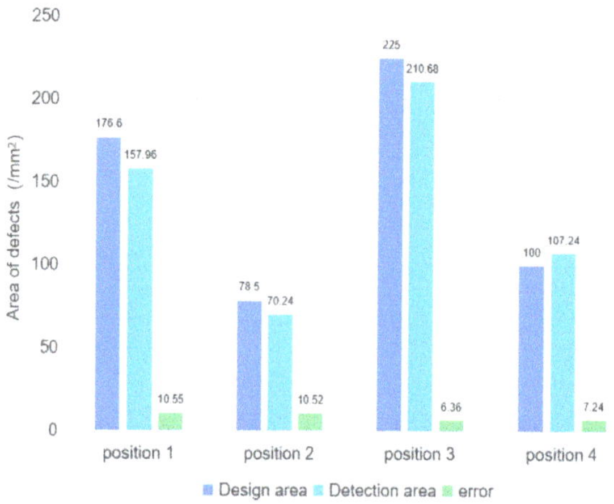

Figure 13. Area detection results.

From the characterization results, it is evident that the adaptive SVD clutter suppression process has greatly improved the accuracy of defect detection. The position detection error is within 0.5 mm, and the area detection error is within 11%, indicating a high level of accuracy.

However, some errors were observed during the characterization process. These errors can be attributed to two main reasons. Firstly, there were processing errors in the specimen defect itself. During the preparation of the specimen, inaccuracies in the cutting size of the polyethylene film caused the actual area of the buried defect to deviate from the designed area. This led to errors in the characterization of the defect area. Secondly, experimental errors were introduced due to variations in the lifting-off of the waveguide probe at different locations on the GFRP specimen. These variations, combined with the inability to guarantee absolute perpendicularity between the specimen and the probe, resulted in inaccurate experimental sampling data.

4. Conclusions

In this paper, we proposed a nondestructive quantitative detection method for internal defects in GFRP. Firstly, the proposed method exhibited excellent imaging performance and effectively highlighted the features of defects. By comparing the 2D images before and after applying the improved clutter suppression method, the effectiveness of the proposed adaptive SVD clutter suppression method and spectral reconstruction algorithm was demonstrated. In addition, the proposed adaptive SVD clutter suppression method can be applied to detect other defects in GFRP using microwave detection. Secondly, the proposed method demonstrated high accuracy in quantitative evaluation. The quantitative detection results demonstrate that the quantitative characterization method is capable of defect detection without leakage, position detection with an error within 0.5 mm, and area detection with an error within 11%.

In summary, the proposed method enables high-quality visual imaging and precise quantitative characterization. Future endeavors will focus on visualizing and quantitatively detecting defects with smaller spatial dimensions.

Author Contributions: Conceptualization, Y.F. and Z.C.; Methodology, Y.F.; Formal analysis, X.Y.; Investigation, R.W.; Data curation, R.W.; Writing—original draft, X.Y.; Writing—review & editing, Y.L.; Supervision, Z.C.; Funding acquisition, Y.L., Y.F. and Z.C. All authors have read and agreed to the published version of the manuscript.

Funding: This work was supported in part by National Magnetic Confinement Fusion Program of China under Grant 2019YFE03130003, the National Natural Science Foundation of China under Grant 11927801, 52107009, Key R&D Project in Shaanxi Province under Grant 2023-YBGY-411.

Institutional Review Board Statement: Not applicable.

Informed Consent Statement: Not applicable.

Data Availability Statement: The data presented in this study are available on request from the corresponding author after obtaining permission of an authorized person.

Conflicts of Interest: The authors declare no conflict of interest.

References

1. Li, Z.; Haigh, A.; Soutis, C. A review of microwave testing of glass fibre reinforced polymer composites. *Nondestruct. Test. Eval.* **2019**, *34*, 429–458. [CrossRef]
2. Meltem, A.; Hasan, G. A review on machinability of carbon fiber reinforced polymer (CFRP) and glass fiber reinforced polymer (GFRP) composite materials. *Def. Technol.* **2018**, *14*, 318–326.
3. Soutis, C. Fibre reinforced composites in aircraft construction. *Prog. Aerosp. Sci.* **2005**, *41*, 143–151. [CrossRef]
4. Liu, D.F.; Tang, Y.J.; Cong, W.L. A review of mechanical drilling for composite laminates. *Compos. Struct.* **2012**, *94*, 1265–1279. [CrossRef]
5. Geng, D.X.; Liu, Y.H.; Shao, Z.Y. Delamination formation, evaluation and suppression during drilling of composite laminates: A review. *Compos. Struct.* **2019**, *216*, 168–186. [CrossRef]
6. Diamanti, K.; Hodgkinson, J.M.; Soutis, C. Detection of low-velocity impact damage in composite plates using lamb waves. *Struct. Health Monit.* **2004**, *3*, 33–41. [CrossRef]
7. Giurgiutiu, V.; Soutis, C. Enhanced composites integrity through structural health monitoring. *Appl. Compos. Mater.* **2012**, *19*, 813–829. [CrossRef]
8. Corigliano, P.; Cucinotta, F.; Guglielmino, E. Thermographic analysis during tensile tests and fatigue assessment of S355 steel. *Procedia Struct. Integr.* **2019**, *18*, 280–286. [CrossRef]
9. Yu, B.; Blanc, R.; Soutis, C. Evolution of damage during the fatigue of 3D woven glass-fibre reinforced composites subjected to tension-tension loading observed by time-lapse X-ray tomography. *Compos. Part A Appl. Sci. Manuf.* **2016**, *82*, 279–290. [CrossRef]
10. Mieloszyk, M.; Majewska, K.; Ostachowicz, W. THz spectroscopy application for detection and localisation of water inclusion in glass composite. *Compos. Struct.* **2018**, *192*, 537–544. [CrossRef]
11. Guerboukha, H.; Nallappan, K.; Skorobogatiy, M. Toward real-time terahertz imaging. *Adv. Opt. Photonics* **2019**, *10*, 843–938. [CrossRef]
12. Zhong, S. Progress in terahertz nondestructive testing: A review. *Front. Mech. Eng.* **2019**, *14*, 273–281. [CrossRef]
13. Li, Z.; Haigh, A.; Soutis, C. Principles and Applications of Microwave Testing for Woven and Non-Woven Carbon Fibre-Reinforced Polymer Composites: A Topical Review. *Appl. Compos. Mater.* **2021**, *25*, 965–982. [CrossRef]
14. Nawaf, S.; Muhammad, F.A. Prospect of Using Artificial Intelligence for Microwave Nondestructive Testing Technique: A Review. *IEEE Access* **2019**, *7*, 110628–110650.
15. Sutthaweekul, R.; Tian, G.Y.; Wang, Z.; Ciampa, F. Microwave open-ended waveguide for detection and characterisation of FBHs in coated GFRP pipes. *Compos. Struct.* **2019**, *225*, 111080. [CrossRef]
16. Liu, C.; Qaseer, M.T.; Zoughi, R. Permittivity Extraction from Synthetic Aperture Radar (SAR) Images of Multilayered Media. *IEEE Trans. Instrum. Meas.* **2021**, *70*, 8004611. [CrossRef]
17. Reza, K.A.; Maryam, R.; Raveena, S. Nondestructive Testing of Nonmetallic Pipes Using Wideband Microwave Measurements. *IEEE Trans. Microw. Theory Tech.* **2020**, *68*, 1763–1772.
18. Akbar, M.F.; Jawad, G.N.; Rashid, L.D. Nondestructive Evaluation of Coatings Delamination Using Microwave Time Domain Reflectometry Technique. *IEEE Access* **2020**, *8*, 114833–114841. [CrossRef]

19. Tan, S.Y.; Akbar, M.F.; Shrifan, N.H.M.M.; Nihad Jawad, G. Assessment of Defects under Insulation Using K-Medoids Clustering Algorithm-Based Microwave Nondestructive Testing. *Coatings* **2022**, *12*, 1440. [CrossRef]
20. Shrifan, N.H.M.M.; Jawad, G.N.; Isa, N.A.M.; Akbar, M.F. Microwave Nondestructive Testing for Defect Detection in Composites Based on K-Means Clustering Algorithm. *IEEE Access* **2021**, *9*, 4820–4828. [CrossRef]
21. Shah, M.B.; Gao, Y.; Ravan, M. Quantitative Defect Size Evaluation in Fluid-Carrying Nonmetallic Pipes. *IEEE Trans. Microw. Theory Tech.* **2022**, *70*, 4071–4081. [CrossRef]
22. Mazzinghi, A.; Freni, A.; Capineri, L. A microwave non-destructive testing method for controlling polymeric coating of metal layers in industrial products. *NDT&E Int.* **2019**, *102*, 207–217.
23. Wang, P.Y.; Li, Z.C.; Pei, Y.M. Microwave nondestructive detection and quantitative evaluation of kissing defects in GFRP laminates. *Compos. Sci. Technol.* **2018**, *162*, 117–122. [CrossRef]
24. Atsushi, H.; Yuhei, Y.; Yang, J. Detection and quantitative evaluation of defects in glass fiber reinforced. *Compos. Struct.* **2015**, *128*, 134–144.
25. Sobkiewicz, P.; Bienkowski, P.; Blazejewski, W. Microwave defectoscopy–detection of composite delamination. *J. Phys. Conf. Ser.* **2021**, *1782*, 012034. [CrossRef]
26. Wang, J.P.; Aubry, P.; Yarovoy, A. 3D Short-Range Imaging with Irregular MIMO Arrays Using NUFFT-Based Range Migration Algorithm. *IEEE Trans. Geosci. Remote Sens.* **2020**, *58*, 4730–4742. [CrossRef]
27. Zhao, M.; Jia, X. A novel strategy for signal denoising using reweighted SVD and its applications to weak fault feature enhancement of rotating machinery. *Mech. Syst. Signal Process.* **2017**, *94*, 129–147. [CrossRef]
28. Le, C.; Dogaru, T.; Nguyen, L. Ultrawideband (UWB) radar imaging of building interior: Measurements and predictions. *IEEE Trans. Geosci. Remote Sens.* **2009**, *47*, 1409–1420. [CrossRef]
29. Zhou, W.H.; Lu, J.P.; Zhou, O. Evaluation of Back Projection Methods for Breast Tomosynthesis Image Reconstruction. *J. Digit. Imaging* **2015**, *28*, 338–345. [CrossRef]
30. Zhang, Y.; Deng, B.; Yang, Q. Near-Field Three-Dimensional Planar Millimeter-Wave Holographic Imaging by Using Frequency Scaling Algorithm. *Sensors* **2017**, *17*, 2438. [CrossRef]
31. Ravan, M.; Amineh, R.K. Two-dimensional near-field microwave holography. *Inverse Probl.* **2010**, *26*, 055011. [CrossRef]
32. Chen, M.; Qiu, X.; Fu, K. Analysis and Compensation for Systematical Errors in Airborne Microwave Photonic SAR Imaging by 2D Autofocus. *IEEE J. Sel. Top. Appl. Earth Obs. Remote Sens.* **2023**, *16*, 2221–2236. [CrossRef]
33. Kou, L.; Wang, X.; Chonq, J. Circular SAR processing using an improved omega-k type algorithm. *J. Syst. Eng. Electron.* **2010**, *21*, 572–579. [CrossRef]
34. Li, N.; Lv, X.; Xu, S. An Improved Water Surface Images Segmentation Algorithm Based on the Otsu Method. *J. Circuits Syst. Comput.* **2020**, *29*, 2050251. [CrossRef]
35. Huang, L.M.; Du, P.Y.; Chen, L. Development of a near-infrared fluorescence imaging system based on fluorescence properties of methylene blue. *J. South. Med. Univ.* **2018**, *38*, 414–420.

sensors

MDPI

Article

A Novel Defect Detection Method for Overhead Ground Wire

Yao Xiao [1], Lan Xiong [1], Zhanlong Zhang [1,*] and Yihua Dan [2]

[1] School of Electrical Engineering, Chongqing University, Chongqing 400044, China; 202111131249@stu.cqu.edu.cn (Y.X.); lxiong@cqu.edu.cn (L.X.)
[2] Department of Electronic Engineering, Tsinghua University, Beijing 100084, China; microdanyihua@outlook.com
* Correspondence: zhangzl@cqu.edu.cn

Abstract: Overhead ground wires typically have strong axial tension and are prone to structural defects caused by corrosion and lightning strikes, which could lead to serious safety hazards. Therefore, it is important to detect defects accurately and quickly to avoid those problems. Existing defect detection methods for overhead ground wires are mainly traditional metal defect detection methods, including eddy current detection, ultrasonic detection, and manual visual inspection. However, those methods have problems of low detection efficiency, high environmental requirements, and insufficient reliability. To solve the above problems, this paper studies a novel type of defect detection technology for overhead ground wire. Firstly, the magnetic leakage characteristics around the defects of overhead ground wires are analyzed, and the defect detection device is designed. Then, the influence of air gap, lift-off distance, defect width, and cross-sectional loss rate on the magnetic flux leakage signal is studied, a novel defect detection method for overhead ground wire is proposed, and experimental verification is carried out. The results show that the proposed method can accurately locate and quantify the defect, which has the advantages of good reliability and high efficiency and lays the foundation for preventing accidents caused by defective overhead ground wires.

Keywords: overhead ground wire; defect; magnetic flux leakage signal; detection device

1. Introduction

The overhead ground wire is mainly used to reduce lightning overvoltage and sometimes for data transmission, which is an important part of the power system. Due to the complex and variable operating environment of overhead ground wire, the probability of a defect in the overhead ground wire structure is extremely high [1–3]. The axial tension of the overhead ground wire structure will have a large stress concentration at the defects' places. If the defects cannot be found and repaired in time, it will eventually lead to the fracture and fall of the overhead ground wire, resulting in great economic loss and serious electric accidents [4–6].

In general, the existing detection methods for overhead ground wires still mainly rely on manual visual inspection methods [7]. This method has problems such as high randomness, low efficiency, etc. Overhead ground wire is a ferromagnetic material. The current detection methods for defects in ferromagnetic materials mainly include magnetic particle flaw detection, eddy current detection, ultrasonic detection, X-ray detection, thermal imaging detection, magnetic leakage detection, and so on [8]. Magnetic particle flaw detection technology is commonly used in the crack detection of aviation products, trains, and ships. The magnetic powder sprayed on the ferromagnetic material will gather in the defective part, and then the cracks are identified using manual inspection or machine vision [9]. The method has high requirements for both the environment and the inspector's skills. The eddy current detection method determines the characteristics of the defects by analyzing the magnetic field changes caused by the eddy currents. Eddy current detection has a low detection depth and is susceptible to environments, which makes it difficult to realize higher precision measurements [10–12]. Magnetic particle and eddy current detection can only detect defects near the surface of the overhead ground wire; ultrasonic detection methods can effectively detect internal defects in the metal structure. Ultrasonic detection methods determine the defects by analyzing ultrasonic waves reflected in defective areas [13,14]. However, ultrasonic detection requires surface treatment, and the signal is also prone to interference, reducing the accuracy of the detection results. X-ray detection technology detects the object by light energy irradiation and penetration of the object, but the hazards of radiation to the human body limit the wide application of this technology [15]. The thermal imaging detection method determines the faults of electric equipment and ferromagnetic materials by detecting abnormal heating. Induced leakage current would heat the defective place of the overhead ground wire, and the thermal image of the overhead ground wire can be collected to judge the defective situation [16]. The reliability of this method is low due to the small abnormal temperature rise at the defects of the overhead ground wire. In summary, the existing manual visual inspection method has problems such as high randomness and subjectivity. Most of the existing methods for detecting defects in ferromagnetic materials have problems such as low detection efficiency, low precision, high detection requirements, lack of reliability, radiation risk, etc. Therefore, a scientific and standardized defect detection method for overhead ground wires is important for the power sector.

At present, the leakage magnetic detection method for steel wire ropes presents the advantages of high efficiency and reliability, and the overhead ground wire and steel wire ropes have similar structural characteristics and material properties, so the leakage magnetic detection method has a huge potential in the field of defect detection of overhead ground wire [17,18]. Refs. [19,20] studied the influence of magnet size, magnetic pole spacing, and lift-off distance on the magnetic leakage signal by the element simulation method. A simple magnetic leakage detection device was designed. However, the recognition accuracy of the device needs to be improved. To accurately and quantitatively detect the degree of wire rope damage, Ref. [21] selects a higher precision and a greater number of sensors. In terms of algorithms, Refs. [22–24] introduced wavelet transform, deep learning, convolution, and other algorithms into the quantitative detection of wire rope damage and achieved certain results, but still cannot fully meet the quantitative detection requirements of wire rope. To solve the current problems in the detection of defects in overhead ground wires, this paper studies the defect detection technology of overhead ground wires based on the principle of magnetic leakage detection. This paper mainly includes the design of the device, the analysis of the forward calculation, and the inverse calculation of the size of the defects. Firstly, the principle of magnetic leakage detection is analyzed, and a device suitable for defect detection for overhead ground wire is designed. Then, the influencing factors of magnetic leakage defect detection for overhead ground wire are analyzed based on the finite element model, and the change rules of defect size and magnetic signal are studied. Finally, the defect detection method for overhead ground wire is proposed and verified by experiments.

2. Defect Detection Theory of Overhead Ground Wire

2.1. The Basic Principle of Detecting the Defect of Overhead Ground Wire

The magnetic flux leakage defect detection principle for overhead ground wires is depicted in Figure 1. A permanent magnet, yoke, air gap, and the overhead ground wire form a magnetic flux circuit, which magnetizes the overhead ground wire axially. When the overhead ground wire is not damaged, the magnetic flux lines are mostly constrained within the wire, and the background magnetic field in the air is very weak. However, when a defect occurs in the overhead ground wire, there is an increase in magnetic resistance at the defect site, resulting in distortion in the magnetic field. The original magnetic flux lines cannot entirely pass through the wire at the defect location, causing some of the magnetic flux to leak from the defect area and form a magnetic leakage field. Figures 2a and 2b, respectively, show the distribution of the magnetic field axial component B_x and the magnetic field radial component B_z along the dotted line in Figure 1b. It can be seen from the figure that the magnetic leakage field signal has obvious distortion at the defect: B_x has a maximum value at the defect. The maximum and minimum values of B_z appear at both ends of the defect, and the value of B_z crosses zero at the center of the defect. The characteristics of magnetic leakage field signals B_x and B_z are closely related to the defect size. The characteristics of the magnetic leakage field are extracted by magnetic sensor measurement, and combined with the signal inversion method, the defect size of overhead ground wire can be accurately detected.

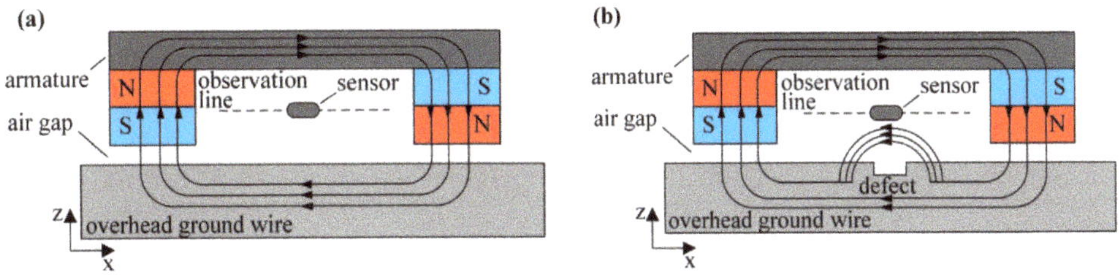

Figure 1. Overhead ground wire magnetic flux leakage detection schematic diagram: (**a**) Without defect; (**b**) Defective.

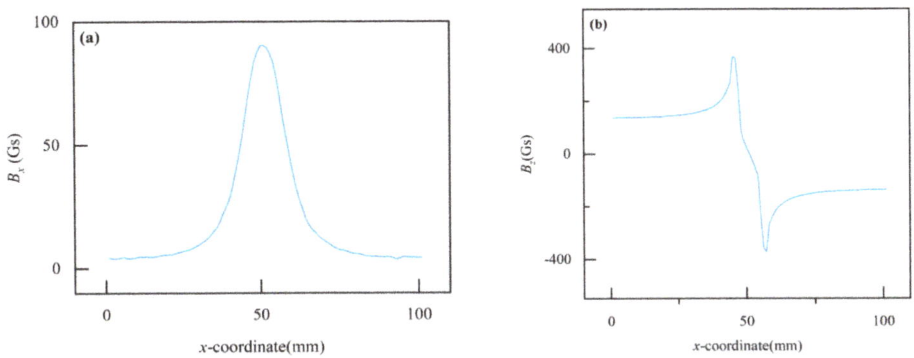

Figure 2. Magnetic flux leakage characteristic signal at the defect: (**a**) B_x; (**b**) B_z.

2.2. Finite Element Calculation Theory of Leakage Magnetic of Overhead Ground Wire

Based on Maxwell's system of equations, the leakage magnetic field distribution of the overhead ground wire in Figure 1 can be described by Equations (1)–(4):

$$\nabla \times \boldsymbol{H} = \boldsymbol{J} \tag{1}$$

$$\nabla \cdot \boldsymbol{B} = 0 \tag{2}$$

$$\boldsymbol{B} = \mu_0 \mu_r \boldsymbol{H} \tag{3}$$

$$\boldsymbol{B} = \nabla \times \boldsymbol{A} \tag{4}$$

where H is the magnetic field strength vector; J is the current density vector; B is the magnetic flux density vector; μ_0 is the vacuum permeability; μ_r is the relative permeability of the overhead ground wire; A is the vector magnetic potential.

Equation (5) is obtained from Equations (1)–(4):

$$\nabla \times \left(\frac{1}{\mu_r \mu_0} \nabla \times A \right) = J \quad (5)$$

Then, the solution region and boundary conditions are set, and Equation (5) is solved numerically by iterative finite element analysis. Finally, the leakage magnetic field distribution of the defective overhead ground wire is calculated.

3. Overhead Ground Wire Defect Detection Device Design

Design the magnetic leakage defect detection device for overhead ground wire, as shown in Figure 3 below. The workflow of the overhead ground wire magnetic leakage detection system is depicted in Figure 4. The functions implemented by the system are modularized, and the main modules in the hardware circuit include the microcontroller control module, AD acquisition module, data storage module, signal conditioning module, wireless transmission module, and more.

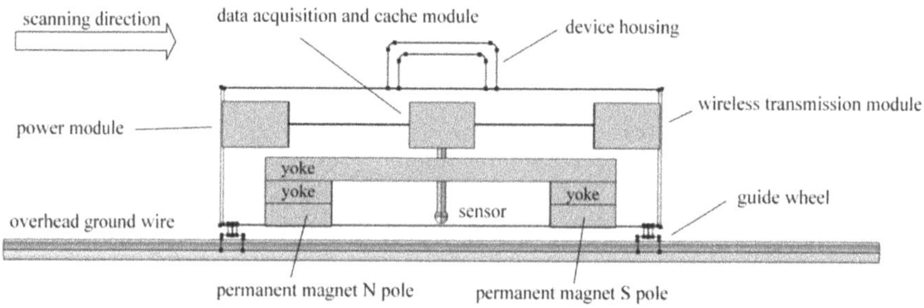

Figure 3. Structural design schematic diagram of the device.

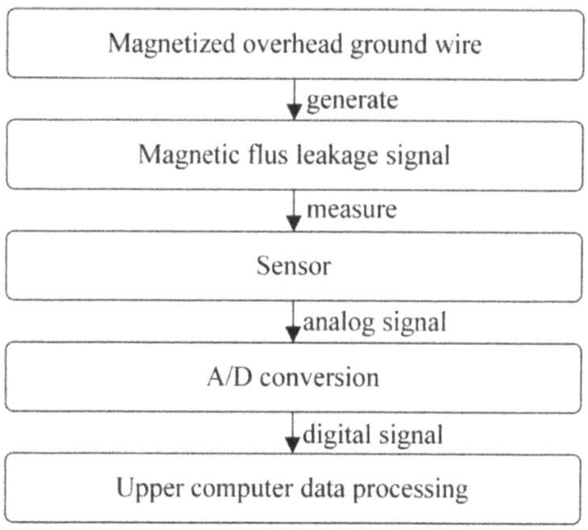

Figure 4. System workflow.

The design of an excitation device holds a crucial position in overhead ground wire defect detection. The saturation magnetization of the overhead ground wire is a prerequisite for accurate defect detection, which can significantly affect the precision of the detection process. Magnetizing the overhead ground wire to saturation results in leakage magnetic signals that are unaffected by material properties, providing stability and accuracy. To reduce the cost of the detection device and ensure its compact structure for portability, the integration of sensors and the excitation structure into a unit has been employed. The permanent magnet NdFeB52 is used to locally axially magnetize the overhead ground wire, and the yoke is used as the magnetic conducting path, further enhancing the magnetization effect. Apart from the excitation device, the air gap between the permanent magnet and the surface of the overhead ground wire, as well as the lift-off distance between the sensor and the surface of the overhead ground wire, are critical factors affecting the accuracy of the detection. This device allows for independent adjustment of these factors through a screw-adjustable height mechanism, optimizing the measurement performance.

The selection of magnetic-sensitive components is the key point of sensor probe design. Hall elements and coils are commonly used in magnetic flux leakage detection. Coils have larger physical dimensions, which are not conducive to compact device design, and they have slower response times. Therefore, the Hall sensor model HG-166A from AsahiKASEI Corporation (Tokyo, Japan) is chosen as the magnetic field sensor. This sensor is characterized by its compact size, rapid response, low power consumption, and high reliability. In practical measurements, it is necessary to simultaneously measure two magnetic field components in two different directions, each with a magnitude of several hundred Gauss. Three-axis Hall sensors can fulfill the requirement of simultaneously detecting two directions of the magnetic field but may not meet the full measurement range requirements. Therefore, two Hall sensors are arranged vertically to measure the magnetic field components in two directions simultaneously. The voltage output range of the Hall sensors is ± 2.5 V. During detection, the device's movement speed is relatively low, and the sampling rate of the analog-to-digital converter (ADC) is not stringent. Hence, the measurement requirements are met by employing the ADS1262 analog-to-digital converter from TEXAS INSTRUMENTS (Dallas, TX, USA). The entire system's data acquisition, storage, and transmission functions are controlled by the STM32F103RCT6 microcontroller (Geneva, Switzerland).

4. Analysis of Magnetic Flux Leakage Distribution Characteristics

The air gap and lift-off distance have a great influence on the magnetic leakage signal. To ensure the stability and reliability of the measurement signal, it is necessary to set a reasonable air gap and lift-off distance before measurement. Therefore, a finite element model is established to analyze the influence of air gap and lift-off distance on the magnetization degree and magnetic leakage signal of the overhead ground wire and to determine the reasonable setting range of air gap and lift-off distance. Then, analyze the change rule of defect size and magnetic leakage signal, and propose the detection method of defect size of overhead ground wire.

4.1. Finite Element Modeling for Magnetic Leakage Analysis of Overhead Ground Wire

References [25,26] indicate that, considering factors such as geometric modeling, mesh partitioning, and computation time, and given that the multi-strand structure of overhead wire has a negligible impact on the leakage magnetic signal at defect locations, a finite element model can be established using a smooth cylindrical equivalent representation for practical overhead wire. Therefore, this paper establishes a finite element model for magnetic leakage analysis of overhead ground wire, as shown in Figure 5 below. Where δ represents the air gap, f represents the lift-off distance, which is the distance between the magnetic field sensor and the surface of the overhead ground wire. The diameter of the overhead ground wire is 9 mm, and the length is 1000 mm.

Due to the influence of different factors and environments, various types of defects, such as broken strands, abrasion, corrosion, fatigue, thermal damage from lightning strikes, etc., can occur on overhead ground wires. These defects have different manifestations, and they may occur simultaneously and interact with each other. Therefore, it is necessary to quantify the size of defects using a uniform standard. The reduced cross-sectional area is the direct cause of the decline in the bearing capacity of the overhead ground wire, so it is reasonable to use the defect width and cross-sectional loss rate to define the size of the defect. The schematic diagram of the defect is shown in Figure 6 below; the location of the defect is at $x = 500$ mm, and the cross-sectional area of the defect is bowed. a represents the width of the defect, S_1 represents the area lost at the defect, and S_2 represents the remaining area at the defect, and the cross-sectional loss rate of the overhead ground wire is η_s, which is calculated by the following equation:

$$\eta_s = \frac{S_1}{S_1 + S_2} \times 100\% \quad (6)$$

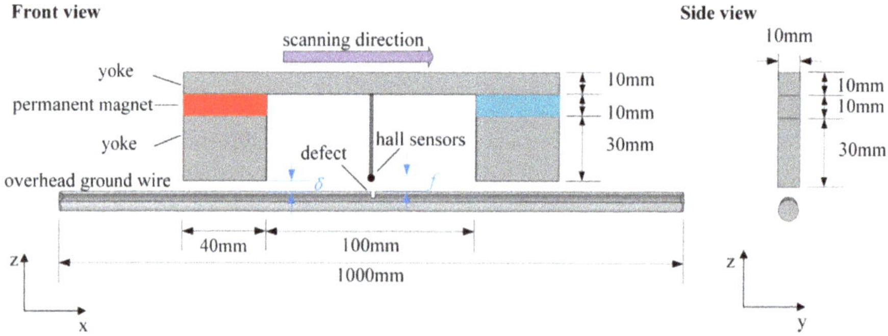

Figure 5. Overhead ground wire simulation calculation model and parameter diagram.

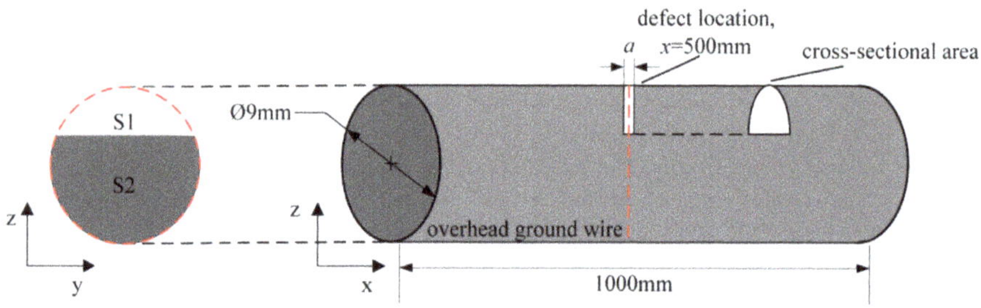

Figure 6. Overhead ground wire defect diagram.

Based on the magnetic properties of the materials, appropriate constitutive relationships are chosen for different materials. The constitutive relationship for the air domain and ferrous materials is represented by the relative magnetic permeability, with values of 1 and 4000, respectively. The magnetic properties of the permanent magnet are expressed in terms of the remanent flux density, which has a value of 1.5 T. The constitutive relationship for the overhead ground wire utilizes a B-H curve, as shown in Figure 7.

After setting appropriate control equations and boundary conditions, calculations are performed using the Magnetic Field module within the AC/DC interface in COMSOL. Finally, the solution of the finite element model of the leakage magnetic field of the overhead ground wire is obtained.

The air gap δ affects the degree of magnetization of the overhead ground wire and, thus, the measurement signal. The lift-off distance f, defect width a, and cross-sectional loss rate η_s also significantly affect the measurement signal. Therefore, the effects of air gap δ, lift-off distance f, defect width a, and cross-sectional loss rate η_s on the measurement signal are investigated in Sections 4.2 and 4.3, respectively.

4.2. *Effect of Air Gap on Magnetic Leakage Signal*

The air gap δ affects the degree of magnetization of the overhead ground wire. Uniform and saturated magnetization of defect-free overhead ground wires by the detection device is required for a stable and strong measurement signal. When δ = 5 mm, the magnetic field distribution across the cross-sectional area of the overhead ground wire without defect is shown in Figure 8. In this figure, it can be observed that two permanent magnets create a closed magnetic flux path, confining almost all magnetic field lines within the overhead ground wire. The magnetic field distribution in the magnetized region of the overhead ground wire is uniform, with a magnetic field strength of approximately 2 T. The region achieves magnetic saturation. Figure 9 shows the relationship between the magnetization level of the overhead ground wire and the air gap δ. Smaller air gaps result in higher magnetization levels. When the air gap increases from 5 mm to 30 mm, the magnetization level of the overhead ground wire decreases from 1.8 T to 1.55 T, a decrease of 0.25 T.

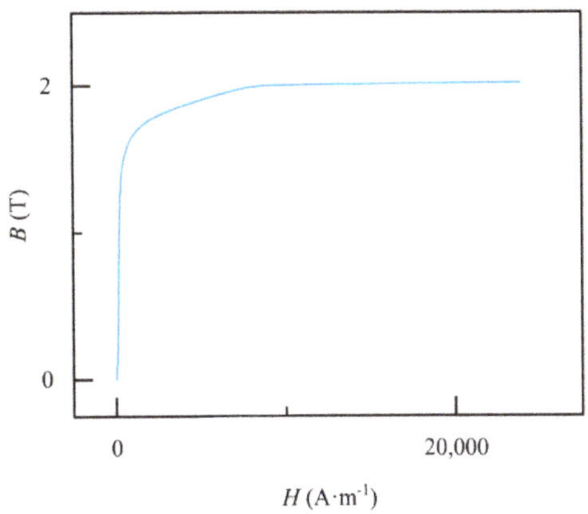

Figure 7. Overhead ground wire B−H curve.

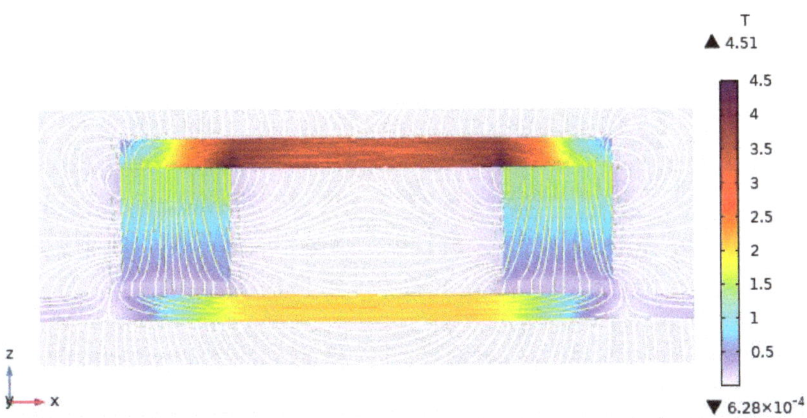

Figure 8. Magnetic field distribution cloud diagram of overhead ground wire without defects.

When the overhead ground wire has a defect (a = 5 mm, η_s = 50%), its magnetic field distribution is shown in Figure 10. Magnetic field lines exhibit distortion at the defect site, with magnetic lines leaking into the air above the defect. Extract the value of magnetic leakage intensity B at 5 mm above the defect, and the change of B with the air gap is shown in Figure 11. Smaller air gaps result in higher leakage magnetic field signal strength. As the air gap increases from 5 mm to 30 mm, the intensity of the leakage magnetic signal above the overhead ground wire defect decreases from 380 Gs to 180 Gs, marking a decline of over 50%.

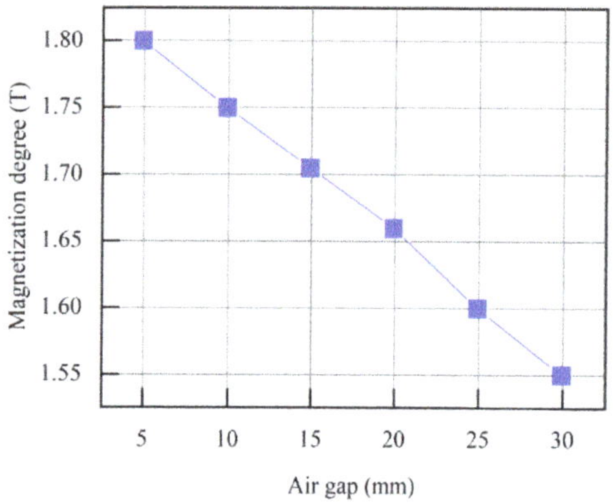

Figure 9. Relationship between the magnetization level of the overhead ground wire and the air gap.

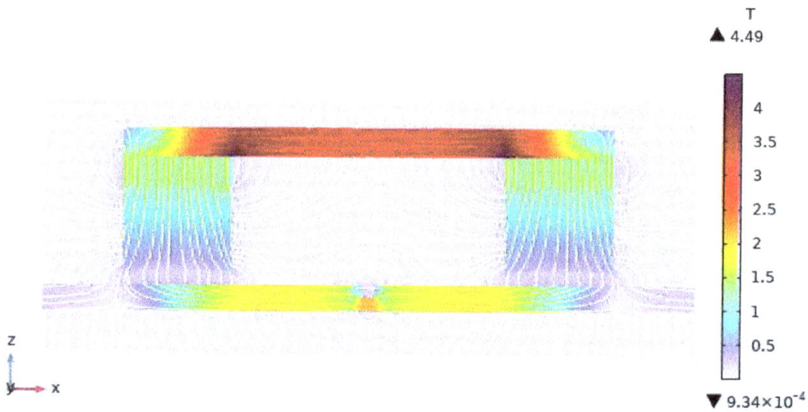

Figure 10. Magnetic field distribution cloud diagram of overhead ground wire with defects.

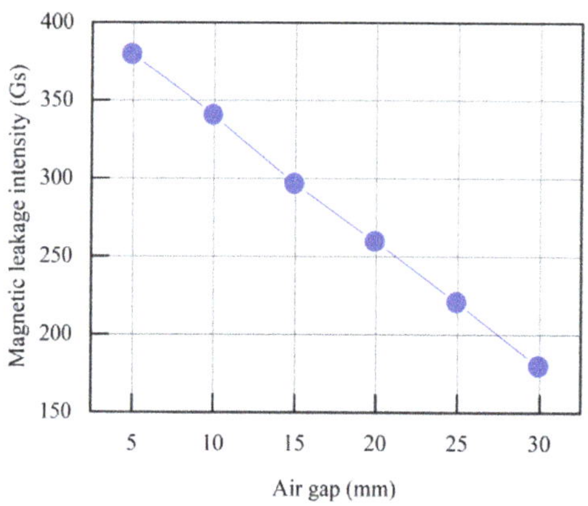

Figure 11. At 5 mm above the defect, there is a relationship between the magnetic flux leakage intensity and the air gap.

In summary, a smaller air gap leads to higher magnetization levels in the overhead ground wire, resulting in stronger leakage magnetic field signals. However, when the gap is too small, the magnetic force between the device and the overhead ground wire is significant, making it less convenient for the device to move along the wire's surface. Therefore, to ensure measurement accuracy and facilitate the movement of the device, the appropriate range for the air gap δ is 5~10 mm.

4.3. Effect of Lift-Off Distance on Magnetic Leakage Signal

The lift-off distance f refers to the distance between the sensor and the surface of the overhead ground wire. To detect the leakage magnetic signal at the defect site effectively, f must be within a certain range; exceeding this range would result in the signal being overshadowed by the background magnetic field. If f is too small, the leakage magnetic signal becomes unstable, making it crucial to select an appropriate lift-off distance.

A vertical spatial magnetic field observation line, as shown in Figure 12, is established directly above the defect, perpendicular to the surface of the overhead ground wire. Magnetic field values are extracted along this observation line under both defect-free and defective conditions, as shown in Figure 13. The horizontal axis, z, ranges from 0 mm (near the surface) to 40 mm (the other end of the observation line). When there is no defect, the magnetic field values along the observation line remain nearly constant, at approximately 100 g. However, when there is a defect, the magnetic field values along the observation line combine with the leakage magnetic field emanating from the defect, resulting in changes in the magnetic field along the observation line. The magnetic field values closer to the surface of the overhead ground wire significantly increase, with the magnitude growing as the observation point moves closer to the wire's surface. Conversely, the magnetic field values further from the surface remain almost the same as the original field, showing minimal variation. To capture the leakage signal containing defect information effectively, the sensor should be positioned within the range of $0 < f < 20$ mm. However, when $f < 5$ mm, the gradient of the magnetic field signal changes substantially, leading to signal instability and reduced noise immunity.

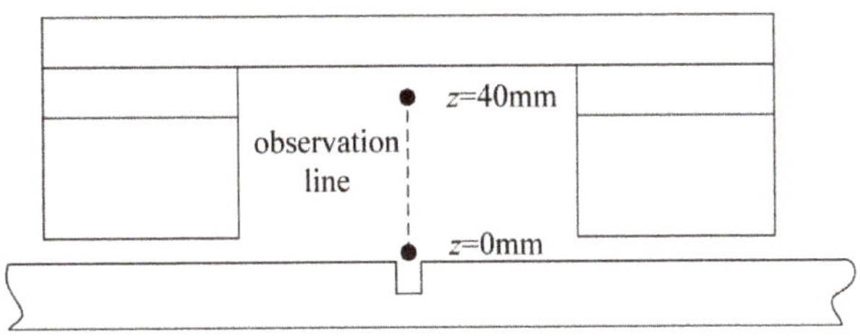

Figure 12. Observation line diagram.

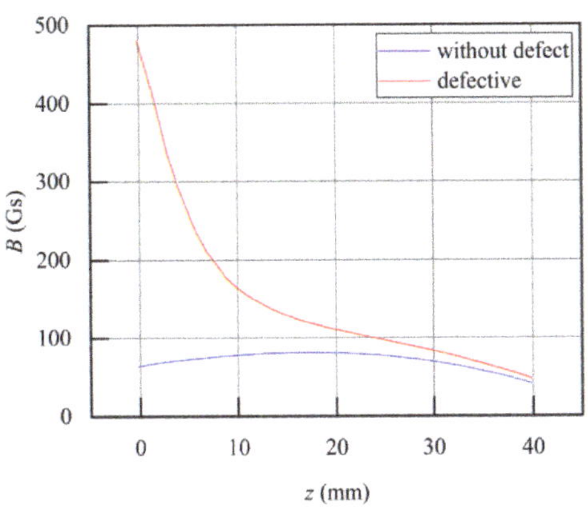

Figure 13. The magnetic field changes on observation lines with and without defects.

4.4. Effect of Defect Size on Magnetic Leakage Signal

With an air gap of δ = 5 mm and a lift-off distance of f = 5 mm in the excitation structure, this paper studies the relationship between the overhead ground wire defects and the leakage magnetic signals. Figure 14a shows the distribution characteristics of the magnetic field signal B_x measured by the Hall sensor when the defect width is a = 10 mm. B_x exhibits peak-like distribution characteristics at the defect site, with maximum $B_{x\text{-max}}$ appearing at the center of the defect. The value of $B_{x\text{-max}}$ increases with the increase in cross-sectional loss. The value of $B_{x\text{-max}}$ presents a monotonic relationship with the magnitude of cross-sectional loss. By conducting several simulations, the distribution characteristics of the detection signal at the defect site were obtained for a range of defect widths from 0.5 mm to 60 mm and cross-sectional loss rate η_s from 1% to 90%. $B_{x\text{-max}}$ values were extracted, and the relationship between $B_{x\text{-max}}$, defect width a, and cross-sectional loss rate η_s is shown in Figure 14b. It can be seen that both defect width and cross-sectional loss rate will significantly influence the magnitude of $B_{x\text{-max}}$. Figure 14c shows the relationship between defect width and $B_{x\text{-max}}$. When the defect width is less than 5 mm, $B_{x\text{-max}}$ increases with an increase in defect width. When the defect width exceeds 5 mm, $B_{x\text{-max}}$ decreases with a further increase in defect width. Figure 14d shows the relationship between the cross-sectional loss rate and $B_{x\text{-max}}$. As the cross-sectional loss rate increases, $B_{x\text{-max}}$ also increases. However, when the cross-sectional loss rate is less than 20%, the change in $B_{x\text{-max}}$ is not substantial.

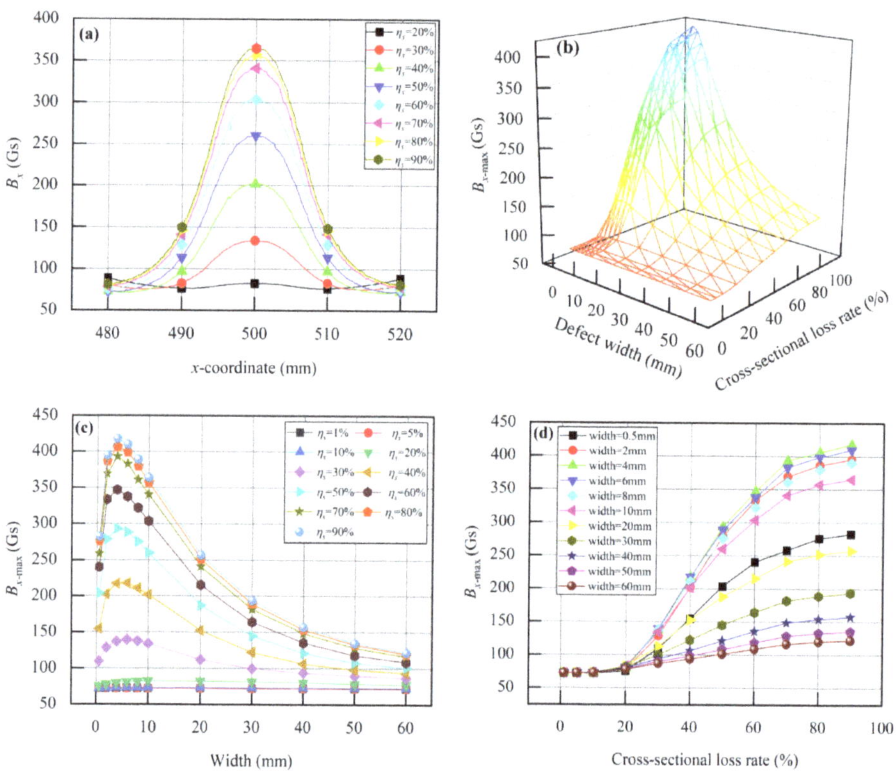

Figure 14. The relationship between defect width, cross-sectional loss rate, and detection signal B_x: (**a**) When the defect width a = 10 mm and the cross-sectional loss rate η_s = 20%~90%, the distribution characteristics of B_x; (**b**) The relationship between $B_{x\text{-max}}$ and defect width a = 0.5~60 mm, cross-sectional loss rate η_s = 1%~90%; (**c**) The relationship between $B_{x\text{-max}}$ and defect width a = 0.5~60 mm; (**d**) The relationship between $B_{x\text{-max}}$ and cross-sectional loss rate η_s = 1%~90%.

Figure 15a shows the distribution characteristics of B_z at the defect site when the defect width is 10 mm. When the cross-sectional loss rate η_s is less than 20%, B_z at the defect site remains relatively constant. However, when the η_s exceeds 20%, B_z exhibits distinctive peak-valley features, with $B_{x\text{-max}}$ increasing as the cross-sectional loss of the overhead ground wire increases. The x-direction spacing between peaks and valleys is denoted as L-B_z, and it is almost unaffected by the cross-sectional loss. Simulation results provide B_z distribution characteristics for various defect widths. When the defect width is less than 2 mm, a B_z does not display peak-valley features. For defect widths greater than or equal to 2 mm, statistical analysis of the relationship between L-B_z and η_s is shown in Figure 15b. The L-B_z increases with increasing defect width. When the defect width is fixed, variations in L-B_z are minimal. Consequently, the value of L-B_z exhibits a strong correlation with the defect width.

In Section 4, through the design of three simulation conditions, the impact of air gap δ, lift-off distance f, defect width a, and cross-sectional loss rate η_s on measurement signals is analyzed. This study reveals that smaller air gaps lead to higher magnetization of the overhead ground wire and stronger leakage of magnetic signals. However, very small distances result in increased magnetic forces between the device and the overhead ground wire, making it less convenient for the device to move along the wire's surface. Excessive lift-off distance f can result in the masking of leakage magnetic signals, and overly small f values may yield unstable signals. The assessment of overhead ground wire defect width and cross-sectional loss rate can be based on comprehensive considerations of L-B_z, $B_{x\text{-max}}$, and $B_{z\text{-max}}$.

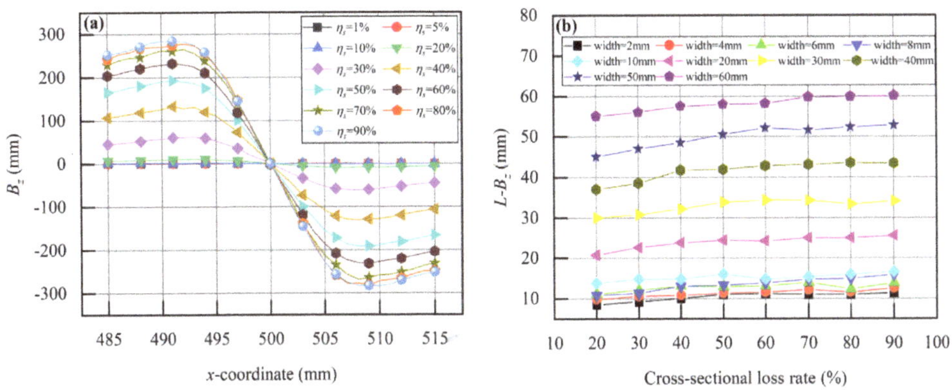

Figure 15. The relationship between defect width and detection signal B_z: (**a**) The distribution characteristics of defect width a = 10 mm, B_z; (**b**) The relationship between the peak-valley spacing L-B_z and the cross-sectional loss rate with different defect widths (2 mm~60 mm).

5. Defect Detection Method for Overhead Ground Wire

Section 4 shows that the peak-to-valley horizontal spacing L-B_z of B_z, the peak $B_{x\text{-max}}$ of B_x, and the peak $B_{z\text{-max}}$ of B_z can be used as parameters for quantitatively determining the size of the defects, but the air gap and the lift-off distance affect the B_x and B_z signals. Once the overall device structure is determined, the magnet's dimensions, magnetic strength, magnetic pole spacing, and other structural parameters remain fixed. Before measurement, it is necessary to optimize the magnetic flux leakage detection effect by adjusting the air gap δ and lift-off distance f. The air gap closely correlates with the magnetization level of the overhead ground wire, favoring a smaller air gap to ensure saturation magnetization. As the distance between the sensor and the overhead ground wire surface increases, the leakage magnetic signal rapidly decreases. Beyond a certain distance, the signal becomes exceedingly weak and is submerged within the background magnetic field, so that the magnetic leakage characteristic signal cannot be detected. Hence, the lift-off distance should not be too large. However, to ensure stable measurement signals, it should not be too small either, necessitating the determination of an appropriate lift-off distance through specific methods. Due to the nonlinear relationship and numerous features between the magnetic leakage characteristic signal and defect size, it is difficult to use fitting functions for defect size inversion. Therefore, following the collection of overhead ground wire magnetic leakage data, a neural network method is employed to perform defect size inversion. Based on the aforementioned analysis, a defect detection method for overhead ground wire is proposed, comprising two main parts: (a) Device design and determination of key structural parameters. (b) Extraction of characteristic parameters and defect size inference

The specific method steps are as follows:

- Step 1: Device design and determination of key structural parameters. The designed structure of the overhead ground wire magnetic flux leakage detection device is shown in Figure 3. Finite element analysis is used to determine a suitable air gap and lift-off distance. Initially, a finite element analysis model is constructed to calculate the relationship between overhead ground wire magnetization intensity B_1 and air gap δ as $B_1 = F_1(\delta)$. When $\delta > \delta_0$, the overhead ground wire has not reached magnetic saturation, whereas when $\delta \leq \delta_0$, the wire has achieved magnetic saturation. Therefore, it is necessary for the air gap δ to be less than or equal to δ_0 to achieve magnetic saturation in the overhead ground wire. Similarly, the relationship between leakage magnetic signal B_2 and lift-off distance f is expressed as $B_2 = F_2(f)$. This relationship between the magnetic leakage signal and lift-off distance is shown in Figure 13. If $f = a$ and $|\partial B_2/\partial f| = 1$, then the lift-off distance f is approximately equal to a.
- Step 2: Magnetic leakage signal acquisition and preprocessing. The overhead ground wire leakage magnetic detection device scans along the wire, with magnetic field sensors collecting data at evenly spaced intervals along the wire's axis. Unlike ideal signals obtained through finite element calculations, practical leakage magnetic signals contain considerable high-frequency noise and data anomalies. Therefore, preprocessing is essential before quantitative analysis. The signal processing workflow, as shown in Figure 16, includes the removal of outlier values using median filtering and the elimination of high-frequency noise using wavelet thresholding.
- Step 3: Extraction of characteristic parameters. The quantitative identification of defect sizes in overhead ground wires is closely related to the extraction of characteristic parameters from the magnetic leakage signals. Therefore, the extraction of characteristics is a crucial step in the analysis of magnetic leakage signals. Based on the rule of how magnetic leakage signals change with defect size, this paper defines the following three magnetic leakage parameters:

1. The peak value of the magnetic leakage signal B_x is denoted as X_1. It represents the maximum value of the magnetic field component B_x at the location of the defect.
2. The peak value of the leakage signal B_z is denoted as X_2. It represents the maximum value of the magnetic field component B_z at the location of the defect.
3. The peak-to-valley horizontal spacing of the magnetic leakage signal B_z is denoted as X_3. It is defined as the distance between the peak and valley of the magnetic leakage signal B_z in the horizontal direction.

- Step 4: Constructing a database of samples. After defining the above characteristic parameters, the magnetic leakage signal on the defective surface of the overhead ground wire can be obtained by using the magnetic dipole computational model, the finite element computational model, and experiments. Then, characteristic values can be extracted to construct a database of magnetic leakage detection samples.
- Step 5: GA-BP (genetic algorithm-back propagation) neural network to predict defect size GA-BP (genetic algorithm-back propagation) neural network to predict defect size. The BP (back propagation) neural network has strong self-adaptation and self-learning abilities and can process large-scale data in parallel. However, the search direction of the BP neural network is relatively simple, which makes the network easy to fall into

the local optimal state. GA (genetic algorithm) is an adaptive parallel optimization algorithm. It simulates the mechanisms of biological genetics and evolution and has the advantages of global search, high parallelism, and strong generalization ability. The GA-BP neural network uses GA to optimize the initial weights and thresholds of the neural network, which are located near the global optimum. Then, the initial weights and thresholds of the neural network are given to the neural network for network training and learning until it converges to find the global optimal solution. The schematic diagram of the BP neural network structure and the specific algorithm flow of the GA-BP neural network are shown in Figures 17 and 18.

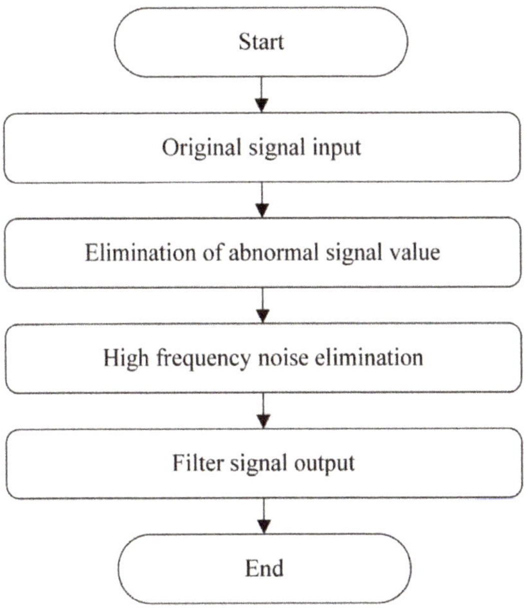

Figure 16. Signal processing flow.

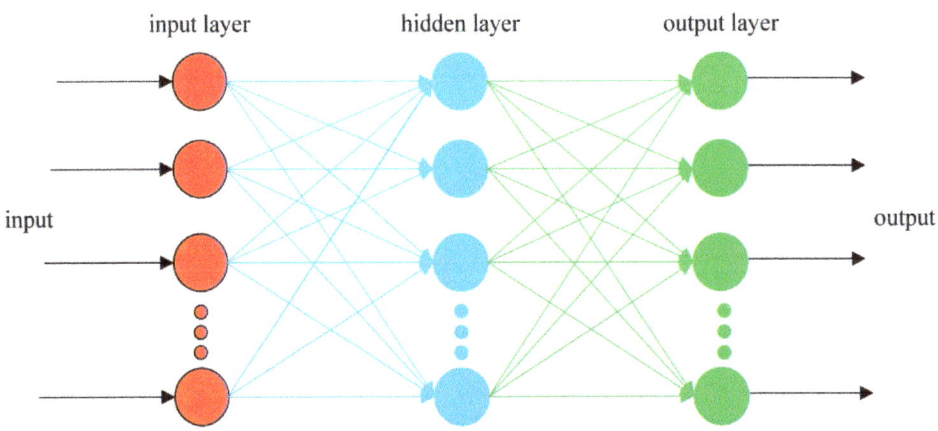

Figure 17. BP neural network structure diagram.

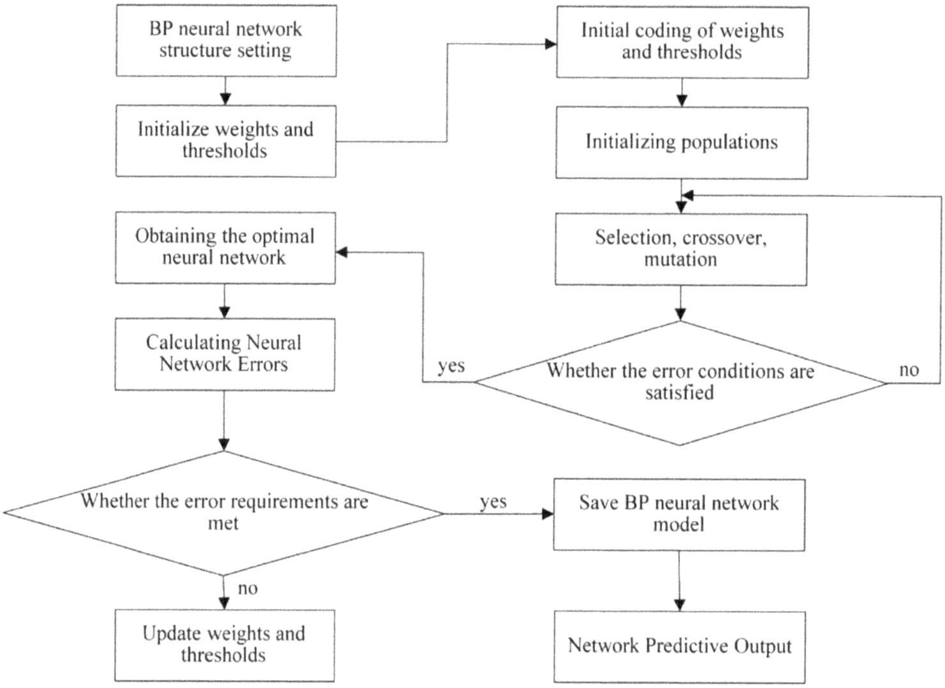

Figure 18. GA-BP neural network training process.

The process of predicting the size of defects in overhead ground wires using the GA-BP neural network is as follows:

4. The sample data are divided into training and test data and normalized. Then, the number of nodes in the input, hidden, and output layers of the BP neural network is determined.
5. Optimize the BP neural network using the genetic algorithm toolbox (GAOT) of Matlab. Save the BP neural network model when the error in the training data meets the objective.
6. Test the network model using test data. Construct a judging metric to judge the prediction accuracy of this network and save the neural network models that predict well.
7. The magnetic leakage signal of the overhead ground wire to be tested is obtained by using the magnetic leakage detection device, and the magnetic leakage characteristic parameters are extracted and brought into the trained neural network model in (c) to obtain the defect size of the overhead ground wire.

6. Experimental Verification

To construct a sample database of real magnetic leakage signals, a simulation experiment platform is built, as shown in Figure 19 below. The dimensional parameters of the device structure are the same as those of the simulation model shown in Figure 3. According to the simulation analysis in Section 2, the air gap δ = 5 mm and the lift-off distance f = 5 mm. The overhead ground wire of model GJ-50 is taken as the object to be detected, the stranded structure is ignored, and the overhead ground wire is simplified as a cylindrical mild steel bar with a diameter of 9 mm. The sensor is a linear Hall sensor, and the magnetic field measurement range is ±500 Gs. The model of the acquisition card is AD7606, 16-bit resolution, with the highest sampling rate of 200 Ksps. The collected data are transmitted to the upper computer through the data line. The telescopic shaft controls the detection scanning speed through unidirectional expansion and contraction, and the speed can be adjusted from 4 mm/s to 30 mm/s. The role of the controller is to control the expansion and contraction of the telescopic axis.

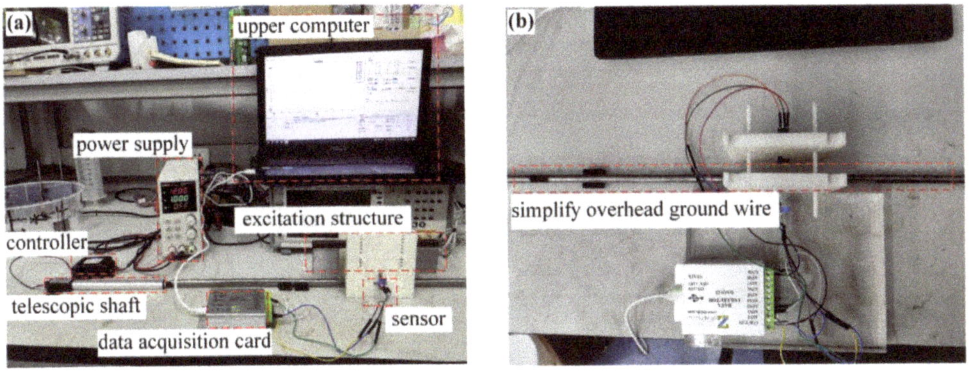

Figure 19. Simulation experiment platform: (**a**) Overall diagram; (**b**) Data acquisition diagram.

Using the wire-cut processing method, 96 samples of overhead ground defect models were produced on simplified overhead ground wires, as shown in Table 1 below.

Table 1. Overhead ground wire defect size parameters.

Sample Number	Cross-Sectional Loss η_s (%)	Defect Width a (mm)	Sample Number	Cross-Sectional Loss η_s (%)	Defect Width a (mm)
1	10	5	49	10	35
2	20	5	⋮	⋮	⋮
3	30	5	56	80	35
4	40	5			
5	50	5	57	10	40
6	60	5	⋮	⋮	⋮
7	70	5	64	80	40
8	80	5			
			65	10	45
9	10	10	⋮	⋮	⋮
⋮	⋮	⋮	72	80	45
16	80	10			
			73	10	50
17	10	15	⋮	⋮	⋮
⋮	⋮	⋮	80	80	50
24	80	15			
			81	10	55
25	10	20	⋮	⋮	⋮
⋮	⋮	⋮	88	80	55
32	80	20			
			89	10	60
33	10	25	⋮	⋮	⋮
⋮	⋮	⋮	96	80	60
40	80	25			
41	10	30			
⋮	⋮	⋮			
48	80	30			

Table 4. *Cont.*

Serial Number	True Values		Predicted Values		Absolute Error	
	Width (mm)	Cross-Sectional Loss (%)	Width (mm)	Cross-Sectional Loss (%)	Width (mm)	Cross-Sectional Loss (%)
13	60	30	60.44312	31.54727	0.443117	1.54727
14	60	40	60.36944	38.92275	0.369444	−1.07725
15	20	40	20.60974	41.01483	0.609742	1.014834
16	45	70	44.34946	69.7722	−0.65054	−0.2278
17	30	60	30.24246	60.71333	0.24246	0.713328
18	55	50	55.27708	50.23921	0.277084	0.239212

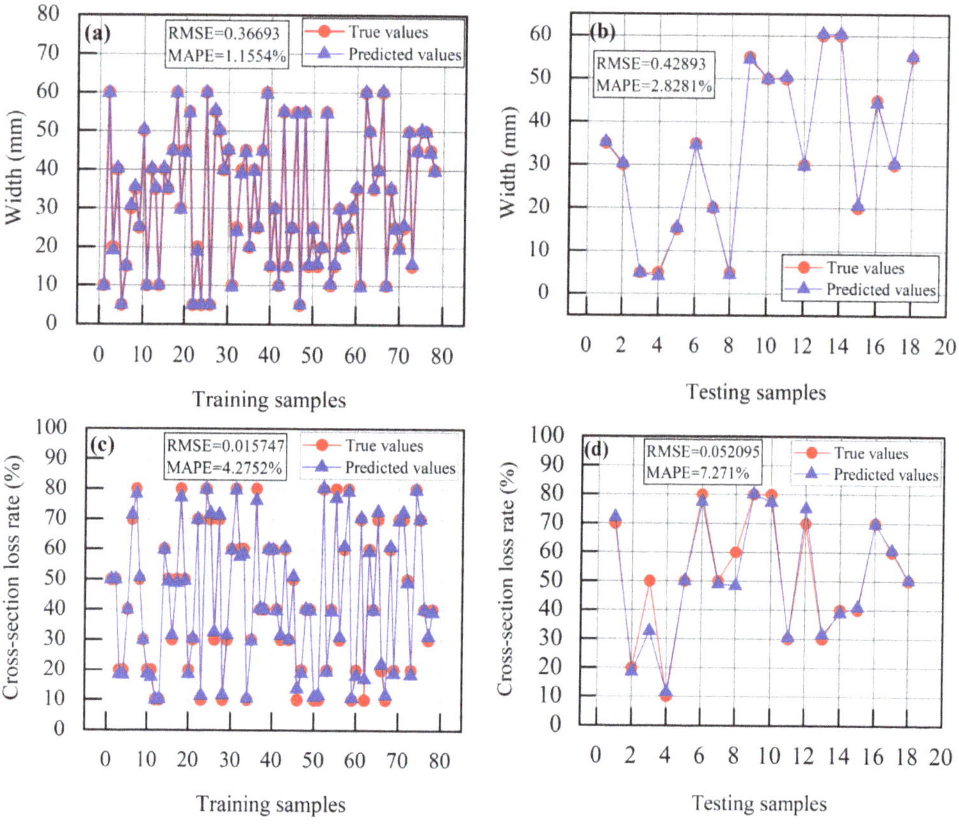

Figure 21. Prediction results of the GA-BP neural network: (**a**) Width prediction training set; (**b**) Width prediction testing set; (**c**) Cross-sectional loss rate prediction training set; (**d**) Cross-sectional loss rate prediction testing set.

7. Conclusions

Overhead ground wires operate in complex environments, suffer long-term corrosion, and are subjected to large axial tensions. Overhead ground wires are prone to structural defects, and further development can lead to serious power safety accidents. Accurate detection of the size of defects in the overhead ground wire can effectively avoid problems caused by structural damage to the overhead ground wire, so accurate detection of the size of defects in the overhead ground wire is essential to ensuring the safe and stable operation of transmission lines. To solve the problems of low detection efficiency, poor reliability, and high environmental requirements of the existing methods, this paper designs a new type of defect detection device for overhead ground wires and proposes a new type of defect detection method for overhead ground wires. The main conclusions are as follows:

8. According to the influence of the air gap and lift-off distance on the magnetic leakage signal, the appropriate air gap range is determined to be 5~10 mm, and the lift-off distance range is 5~20 mm. The smaller the air gap and lifting distance, the stronger the magnetic leakage signal. However, the air gap being too small will result in too much magnetic force, making it difficult to move the device; the lift-off distance being too small will lead to signal instability. Increasing the lift-off distance improves the stability of the signal, but the leakage signal will be gradually covered by the background magnetic field, and the sensitivity will be reduced.
9. B_x, B_z, and L-B_z are determined as three characteristic quantities reflecting the size of defects. The peak values of B_x and B_z gradually increase with the increase in the cross-sectional loss rate of defects, and L-B_z gradually increases with the increase in the width of defects. These three characteristic quantities are closely related to the defect size and can be utilized to evaluate the defect width and the defect cross-sectional loss rate.
10. A defect detection method for overhead ground wire is proposed, which is verified by the experimental tests. The experimental results show that the absolute errors of the predicted value of overhead ground wire width are all within 1 mm, and the absolute errors of the predicted value of overhead ground wire cross-sectional loss rate are within 3% for more than 80% of the samples.

In summary, the proposed method can solve the problems of existing methods and has the advantages of greater measurement depth, no detection blind spots, good environmental adaptability, high detection accuracy, stability, etc. This work can detect defects in overhead ground wires accurately and efficiently, which improves the efficiency of overhead ground wire maintenance and has great engineering significance.

Author Contributions: Conceptualization, Y.X. and L.X.; methodology, Y.X.; software, Y.D.; validation, Y.X., L.X. and Y.D.; formal analysis, Y.X.; investigation, Y.X.; resources, Z.Z.; data curation, Z.Z.; writing—original draft preparation, Y.X.; writing—review and editing, Z.Z.; visualization, L.X.; supervision, Z.Z.; project administration, Z.Z.; funding acquisition, Z.Z. All authors have read and agreed to the published version of the manuscript.

Funding: This research received no external funding.

Institutional Review Board Statement: Not applicable.

Informed Consent Statement: Not applicable.

Data Availability Statement: Data are contained within the article.

Conflicts of Interest: The authors declare no conflict of interest.

References

1. Kondrateva, O.; Myasnikova, E.; Loktionov, O. Analysis of the Climatic Factors Influence on the Overhead Transmission Lines Reliability. *Sci. J. Riga Tech. Univ. Environ. Clim. Technol.* **2021**, *24*, 201–214. [CrossRef]
2. Yang, S.; Zhou, W.; Zhu, S.; Wang, L.; Ye, L.; Xia, X.; Li, H. Failure probability estimation of overhead transmission lines considering the spatial and temporal variation in severe weather. *J. Mod. Power Syst. Clean Energy* **2019**, *7*, 131–138. [CrossRef]
3. Burtsev, A.V.; Selivanov, V.N. Method for Calculating of Lightning Impact Level on Overhead Transmission Line. *Appl. Sci.* **2021**, *11*, 6609. [CrossRef]
4. Liu, G.; Peng, X.; Zhong, M.; Wang, R.; Wang, P.; Guo, D.; Qi, Q. A case study of ruptures in overhead ground wire under a large lightning over 400kA. *Eng. Fail. Anal.* **2019**, *104*, 1211–1233. [CrossRef]
5. Sun, J.; Li, S.; Qing, Q.; Guo, Y.; Yao, X.; Wu, Y. Damage Characteristics of OPGWs under Consecutive Lightning Strikes Based on a Coupled Arc-Thermal-Electric Simulation. *IEEE Trans. Power Deliv.* **2022**, *38*, 1927–1937. [CrossRef]
6. Wang, C.; Wang, S.; Li, J.; Wu, J.; Zhong, J.; Liao, H. Research on the Identification Method of Overhead Transmission Line Breeze Vibration Broken Strands Based on VMD−SSA−SVM. *Electronics* **2022**, *11*, 3028. [CrossRef]

7. Ni, J.; Zhan, T. Smart Sensor Network Optimization and Line Defect Identification Based on UAV Transmission Inspection under 5G Technology. *J. Sens.* **2022**, *2022*, 1–8. [CrossRef]
8. Zhao, L.; Zhang, T.; Huang, X.; Zhang, Y.; Liu, W. A structural health monitoring system of the overhead transmission line conductor. *IET Sci. Meas. Technol.* **2021**, *16*, 28–39. [CrossRef]
9. Ye, J.-H.; Ni, R.-H.; Hsu, Q.-C. Image feature analysis for magnetic particle inspection of forging defects. *Proc. Inst. Mech. Eng. Part B J. Eng. Manuf.* **2021**, *236*, 1923–1929. [CrossRef]
10. AbdAlla, A.N.; A Faraj, M.; Samsuri, F.; Rifai, D.; Ali, K.; Al-Douri, Y. Challenges in improving the performance of eddy current testing: Review. *Meas. Control.* **2018**, *52*, 46–64. [CrossRef]
11. Grochowalski, J.M.; Chady, T. Pulsed Multifrequency Excitation and Spectrogram Eddy Current Testing (PMFES-ECT) for Nondestructive Evaluation of Conducting Materials. *Materials* **2021**, *14*, 5311. [CrossRef] [PubMed]
12. Liao, C.; Yi, Y.; Chen, T.; Cai, C.; Deng, Z.; Song, X.; Lv, C. Detecting Broken Strands in Transmission Lines Based on Pulsed Eddy Current. *Metals* **2022**, *12*, 1014. [CrossRef]
13. Anane, P.O.K.; Huang, Q.; Ayimbire, P.N.; Bamisile, O. Non-contact measurement of traveling wave of overhead transmission line. *Measurement* **2021**, *181*, 109557. [CrossRef]
14. Honarvar, F.; Varvani-Farahani, A. A review of ultrasonic testing applications in additive manufacturing: Defect evaluation, material characterization, and process control. *Ultrasonics* **2020**, *108*, 106227. [CrossRef]
15. Mousa, T.O.; Mohammed, M.S.; Aljohani, M.S. Optimizing Radiographic Sensitivity in the In-Service Testing of Pipes. *Russ. J. Nondestruct. Test.* **2020**, *56*, 92–99. [CrossRef]
16. Jalil, B.; Pascali, M.A.; Leone, G.R.; Martinelli, M.; Moroni, D.; Salvetti, O.; Berton, A. Visible and Infrared Imaging Based Inspection of Power Installation. *Pattern Recognit. Image Anal.* **2019**, *29*, 35–41. [CrossRef]
17. Feng, B.; Wu, J.; Tu, H.; Tang, J.; Kang, Y. A Review of Magnetic Flux Leakage Nondestructive Testing. *Materials* **2022**, *15*, 7362. [CrossRef]
18. Zhang, O.; Wei, X. Analysis of MFL Model for Sucker Rod Defects and Its MFL Signal Processing. *J. Test. Eval.* **2018**, *47*, 3765–3780. [CrossRef]
19. Xu, F.; Jiang, Z.; Jiang, Q.; Wang, X.S. Damage Detection and Assessment of Broken Wires in Cables of a Bridge Based on Magnetic Flux Leakage. *Exp. Tech.* **2022**, *47*, 907–920. [CrossRef]
20. Ding, D.; Lu, J.; Xu, F. Designing and Investigating a Nondestructive Magnetic Flux Leakage Detection System for Quantitatively Identifying Wire Defects. *IEEE Sens. J.* **2022**, *22*, 20360–20372. [CrossRef]
21. Gong, W.; Yang, T.; Guo, L. Magnetic flux leakage detection method by the arrays consisted of three-dimensional hall sensor for rail top surface cracks. *Ferroelectrics* **2022**, *597*, 52–64. [CrossRef]
22. Zhang, J.; Li, J. Recognition of broken wire rope based on adaptive parameterless empirical wavelet transform and random forest. *Insight Non-Destr. Test. Cond. Monit.* **2019**, *61*, 380–394. [CrossRef]
23. Huang, S.; Peng, L.; Sun, H.; Li, S. Deep Learning for Magnetic Flux Leakage Detection and Evaluation of Oil & Gas Pipelines: A Review. *Energies* **2023**, *16*, 1372.
24. Wang, H.; Chen, G. Defect size estimation method for magnetic flux leakage signals using convolutional neural networks. *Insight Non-Destr. Test. Cond. Monit.* **2020**, *62*, 86–91. [CrossRef]
25. Zhou, C.; Wei, C.; Wang, W. A New Detection Method Based on Magnetic Leakage Theory and BP Neural Network for Broken Steel Strands in ACSR Conductor. *IEEE Sens. J.* **2022**, *22*, 19620–19634. [CrossRef]
26. Jiang, X.; Xia, Y.; Hu, J.; Yin, F.; Sun, C.; Xiang, Z. Optimal design of MFL sensor for detecting broken steel strands in overhead power line. *Prog. Electromagn. Res.* **2011**, *121*, 301–315. [CrossRef]

sensors

MDPI

Article

Study on the Ground-Penetrating Radar Response Characteristics of Pavement Voids Based on a Three-Phase Concrete Model

Shuaishuai Wei [1], Huan Zhang [2,*], Jiancun Fu [2] and Wenyang Han [2]

1 School of Civil Engineering, Qingdao University of Technology, Qingdao 266520, China; shuaishuai_wei@163.com
2 Shandong Transportation Institute, Jinan 250102, China; fujiancunhit@163.com (J.F.); hanwenyang@sdjtky.cn (W.H.)
* Correspondence: zhanghuan095@163.com

Abstract

Concrete pavements frequently develop subsurface voids between surface and base layers during long-term service due to cyclic loading, environmental effects, and subgrade instability, which compromise structural integrity and traffic safety. Ground-penetrating radar (GPR) has been widely used as a non-destructive method for detecting such voids. However, the presence of coarse aggregates with strong electromagnetic scattering properties often introduces pseudo-reflection signals in radar images, hindering accurate void identification. To address this challenge, this study develops a high-fidelity three-phase concrete model incorporating aggregates, mortar, and the interfacial transition zone (ITZ). The Finite-Difference Time-Domain (FDTD) method is used to simulate electromagnetic wave propagation in both voided and intact structures. Simulation results reveal that aggregate-induced scattering can blur or distort reflection interfaces, generating pseudo-hyperbolic anomalies even in the absence of voids. In cases of thin-layer voids, real echo signals may be masked by aggregate scattering, leading to missed detections. GPR systems can be broadly classified into impulse, continuous-wave, and multi-frequency types. To validate the simulations, field tests using multi-frequency 2D/3D GPR systems and borehole verification were conducted. The results confirm the consistency between simulated and actual radar anomalies and validate the proposed model. This work provides theoretical insight and modeling strategies to enhance the interpretation accuracy of GPR data for subsurface void detection in concrete pavements.

Keywords: three-phase concrete pavement model; void detection; ground penetrating radar; aggregate scattering; FDTD simulation

1. Introduction

With the rapid advancement of urbanization, road networks in medium and large cities have become increasingly dense, and concrete pavements are widely utilized in municipal roads and critical transportation hubs [1–3]. During service, urban concrete pavements are prone to developing subsurface defects, such as voids and cracks, at the interface between the surface and base layers due to repeated traffic loads and environmental factors, including temperature and moisture variations [4–7]. These voids result from a combination of factors, including insufficient base compaction, underground utility leakage, soil erosion, freeze–thaw cycles, fatigue damage from traffic loads, and differential settlement of the subgrade. Once formed, voids undermine the support relationship between the pavement slab and the base layer, leaving the structure in a state of suspended loading. This not only accelerates water ingress and pumping effects but also intensifies bottom erosion and cracking, ultimately leading to local settlement or even large-scale collapse [8–10]. If such voids are not detected and addressed early, they may rapidly evolve into irreversible structural failures with potentially catastrophic impacts on urban operations. Hence, developing an efficient, accurate, and non-destructive detection and assessment system has become an urgent need for urban pavement maintenance [11].

To date, a variety of physical and geophysical methods have been established for subsurface void detection, including acoustic methods [12–14], Distributed Optical Fiber Vibration Sensing (DOVS) [15], Falling Weight Deflectometers (FWD) [16], Dynamic Cone Penetration tests (DCP) [17], and Ground Penetrating Radar (GPR) [3,18–20]. Although acoustic methods, DOVS, FWD, and DCP can reveal subsurface conditions to some extent, they generally suffer from low efficiency, complex operation, and limited adaptability. GPR, on the other hand, has emerged as a powerful tool for void detection due to its non-contact nature, high resolution, and real-time imaging capabilities [21]. Depending on the type of emitted signal, GPR systems can be generally divided into three classes: (i) impulse radars, which emit short pulses typically similar to a single period of a sinusoid; (ii) continuous-wave radars with varying frequency laws; and (iii) holographic radars, which are usually multi-frequency and designed to produce three-dimensional subsurface images [22–26]. Impulse GPR has been widely employed in road engineering due to its relatively simple system structure and strong adaptability, whereas holographic radars have demonstrated superior imaging resolution and recognition capabilities, enabling fine-scale subsurface imaging even under complex conditions [23–26]. Recent advances in synthetic spectrum imaging and back-projection algorithms further enhance holographic GPR performance, showing high resolution and fast computation [25]. Research on the application of GPR in non-destructive testing (NDT) of roads and concrete structures has yielded substantial progress, covering structural identification, damage detection, parameter inversion, and intelligent data processing. For concrete structures, GPR is widely used in rebar detection, void identification, and structural evaluation due to its high resolution and non-destructive nature. For instance, studies have summarized typical GPR practices for reinforced concrete detection, showing that dual-polarization antennas combined with migration processing significantly improve the accuracy of rebar recognition [27]. Other works have investigated how antenna frequency and void geometry affect GPR signal characteristics, offering guidance for selecting detection parameters [28]. Moreover, the integration of GPR and eddy current testing has revealed the influence of reinforcement layout and concrete cover thickness on crack propagation, demonstrating the technique's potential in analyzing the durability of continuously reinforced concrete pavements [29].

In the context of asphalt concrete pavements, GPR is primarily used to detect density, moisture content, cracks, and the condition of structural layers. Some studies have developed numerical models based on three-phase dielectric theory and validated them through field testing to analyze the relationship between pavement density and dielectric properties [30]. Filtering algorithms have also been proposed to mitigate noise caused by antenna vibrations, thereby enhancing real-time monitoring capabilities during construction [31]. For moisture monitoring, numerical models have been developed under varying saturation levels, with empirical formulas proposed to relate dielectric constant and volumetric water content [32]. In terms of structural integrity assessment, preliminary systems have been established using three-dimensional (3D) GPR for disease feature recognition and the Pavement Internal Condition Index (PICI) [33]. However, the heterogeneous nature of pavement materials and environmental interference limits the applicability of numerical models, and existing filtering techniques often struggle to address strong noise interference.

The development of 3D GPR in recent years has provided a technical breakthrough for high-precision identification of subsurface defects in pavement structures. Unlike traditional two-dimensional (2D) radar, which only captures vertical profile images, 3D radar employs antenna arrays to collect large-scale, high-resolution volumetric data in real-time during vehicle motion, enabling the 3D visualization of structural integrity and damage features. Depending on the coupling method between the antenna and pavement surface, 3D GPR systems are classified into air-coupled and ground-coupled types. Air-coupled antennas are easy to deploy and suitable for rapid, large-area surveys—such as evaluating pavement layer thickness and structural uniformity—but their limited penetration depth and spatial resolution, due to air gaps, hinder the identification of deeper or smaller-scale anomalies. In contrast, ground-coupled antennas enhance signal-to-noise ratio (SNR) by maintaining contact with the surface and supporting higher-frequency operation. They are more suitable for detecting surface cracks, shallow voids, and delamination, and are widely used for high-precision damage localization and assessment [34–36]. Recently, air-coupled horn antennas have been widely adopted for non-destructive electromagnetic testing of pavement structures, primarily for assessing interlayer interfaces [37–39], compaction quality [40,41], moisture content in soil and subgrade [42], and for analyzing dielectric and mechanical properties of materials [43]. However, due to their limited sampling rate, air-coupled antennas exhibit constrained spatial resolution under high-speed operation, making it difficult to identify small-scale or poorly defined defects such as base-layer voids or localized delamination. In contrast, ground-coupled antennas offer higher spatial resolution and SNR, making them better suited for detecting shallow subsurface discontinuities [44]. High-frequency ground-coupled antennas can effectively detect small-scale defects in concrete structures, including cracks, repairs, and localized voids; however, their performance is significantly influenced by material properties, target size, dielectric contrasts, and interface morphology [45]. Therefore, accurately simulating GPR electromagnetic responses to pavement voids requires realistic structural modeling and high-resolution antenna parameters.

In summary, current GPR techniques still face numerous challenges in detecting subsurface defects in concrete pavements, particularly in identifying base-layer voids, filtering pseudo-reflections, and resolving fine, shallow features. Conventional homogeneous models fail to accurately represent the true multiphase structure of concrete, resulting in substantial deviations in wave propagation paths, reflection patterns, and signal interpretations, thereby compromising diagnostic accuracy. To address these issues, this study proposes a multiphase concrete model based on the composition of aggregate, mortar, and interfacial transition zone (ITZ). Various void types and structural parameters are modeled to simulate electromagnetic wave propagation using the Finite-Difference Time-Domain (FDTD) method [46,47]. In this study, the FDTD method was chosen as the core numerical simulation approach for the following key reasons: First, most commercial 2D and 3D GPR systems, such as the 3D-Radar and ImpulseRadar used in this study, perform data acquisition and processing in the time domain. Therefore, employing the FDTD method ensures better alignment with practical engineering applications. Second, the FDTD method directly solves Maxwell's equations and maintains high accuracy and stability, even in complex heterogeneous media, allowing for effective characterization of electromagnetic wave propagation, reflection, and scattering in various geological materials. Additionally, the FDTD method offers the advantage of explicit time-domain computation. It provides complete time-series data. This data not only supports detailed analysis of GPR signal reflections but also lays a solid foundation for subsequent radar waveform feature analysis.

Thus, the adoption of the FDTD algorithm is necessary both to maintain consistency with existing radar systems and to study the electromagnetic response behavior of heterogeneous media. The study systematically investigates how aggregate heterogeneity distorts GPR signals and contributes to the formation of pseudo-anomalies. By comparing B-scan responses from 11 representative models, the research identifies distinct scattering patterns associated with aggregates versus true void features. Furthermore, field data collected using multi-frequency GPR systems are used to validate simulation results. The findings offer theoretical insights and modeling strategies for improving the accuracy of void detection in concrete pavements, thereby opening new avenues for GPR data interpretation and engineering defect diagnosis.

2. FDTD Numerical Simulation

In 2D FDTD simulations [46,47], the most commonly employed configuration is the transverse magnetic (TM) mode, where the electric field is oriented along the z-direction, and the magnetic fields are distributed along the x- and y-directions. This mode is well-suited for modeling wave propagation problems in 2D cross-sectional domains. Assuming that the electromagnetic fields vary only in the x–y plane, with the electric field containing only the E_z component and the magnetic field comprising H_x and H_y, Maxwell's equations can be simplified as follows:

$$\begin{aligned} &\frac{\partial H_y}{\partial x} - \frac{\partial H_x}{\partial y} = \varepsilon \frac{\partial E_z}{\partial t} \\ &\frac{\partial E_z}{\partial y} = -\mu \frac{\partial H_x}{\partial t} \\ &-\frac{\partial E_z}{\partial x} = -\mu \frac{\partial H_y}{\partial t} \end{aligned} \tag{1}$$

where E_z is the out-of-plane electric field component (V/m), H_x and H_y are the magnetic field components along the x and y directions (A/m), ε is the permittivity (F/m), and μ is the permeability (H/m).

The key procedure in the FDTD method is the discretization of the partial differential equations in both the spatial and temporal domains. This involves approximating continuous variables on a discrete grid, enabling numerical time-stepping iteration. To numerically solve Maxwell's equations on a computer, the computational domain must be discretized spatially and temporally.

For 2D GPR problems, the computational domain is typically defined as a rectangular area in the x–y plane, with dimensions $L_x \times L_y$. This region is uniformly discretized into a grid with N_x and N_y nodes along the x and y directions, respectively. The spatial steps are $\Delta x = L_x/N_x$ and $\Delta y = L_y/N_y$, and each grid point is identified by the index (i, j), corresponding to the spatial location $(i \cdot \Delta x,\ j \cdot \Delta y)$. The total simulation time T is divided into N_t steps of size $\Delta t = T/N_t$, and the electromagnetic field values are iteratively updated at each time step n.

To ensure symmetry and numerical stability, the electric and magnetic field components are spatially staggered on a Yee grid [48]. Specifically, the electric field component E_z, which is normal to the x–y plane, is located at the primary grid point (i, j). The magnetic field component H_x is located at $(i, j + 1/2)$, i.e., between the top and bottom of the E_z point, while H_y is located at $(i + 1/2, j)$, i.e., between the left and right of the E_z point. Applying central difference approximations to the spatial and temporal derivatives, the update equations for the TM mode in 2D are derived as:

$$\begin{aligned} &E_z^{n+1}(i,j) = E_z^{n}(i,j) + \frac{\Delta t}{\varepsilon(i,j)}\left[\frac{H_y^{n+1/2}(i,j)-H_y^{n+1/2}(i-1,j)}{\Delta x} - \frac{H_x^{n+1/2}(i,j)-H_x^{n+1/2}(i,j-1)}{\Delta y}\right] \\ &H_x^{n+1/2}(i,j) = H_x^{n-1/2}(i,j) - \frac{\Delta t}{\mu(i,j)} \cdot \frac{E_z^{n}(i,j+1)-E_z^{n}(i,j)}{\Delta x} \\ &H_y^{n+1/2}(i,j) = H_y^{n-1/2}(i,j) + \frac{\Delta t}{\mu(i,j)} \cdot \frac{E_z^{n}(i+1,j)-E_z^{n}(i,j)}{\Delta x} \end{aligned} \tag{2}$$

These equations are iteratively solved over time to simulate the dynamic propagation of electromagnetic waves in the 2D domain. To ensure numerical stability, the FDTD method must satisfy the Courant–Friedrichs–Lewy (CFL) condition [49], which constrains the maximum allowable time step:

$$\Delta t \leq \frac{1}{c\sqrt{\frac{1}{\Delta x^2} + \frac{1}{\Delta y^2}}} \tag{3}$$

where c is the speed of electromagnetic waves in the medium. This condition is closely tied to spatial discretization and the properties of the medium.

In practice, as the computational domain is finite, absorbing boundary conditions must be applied to prevent artificial reflections from the edges. The most widely used method is the perfectly matched layer (PML) [50], which effectively absorbs outgoing waves without reflecting them into the computational region. Unlike earlier absorbing boundary conditions [51], the PML can theoretically achieve reflection-free absorption for electromagnetic waves incident at arbitrary angles and frequencies. This property is crucial for GPR simulations where scattered waves from aggregates and void interfaces impinge on the domain boundaries over a wide range of directions. By using PML, spurious boundary reflections are minimized, ensuring that the recorded signals originate only from the modeled pavement structure. The specific FDTD process is described in Figure 1.

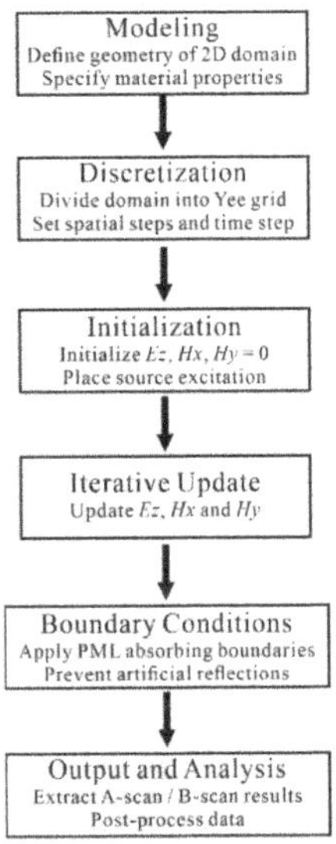

Figure 1. Block diagram of FDTD implementation steps.

3. Concrete Pavement Structure and Modeling

Concrete pavement is a type of rigid pavement widely used in urban roads, highways, airport runways, and industrial infrastructure. Compared with asphalt pavement, it offers several advantages, including higher strength, better durability, lower deformation, and longer maintenance intervals. Concrete pavements are typically composed of multiple layers. Each layer has a distinct function and material composition, working together to carry traffic loads and resist environmental effects. A typical structure consists of three components: a surface layer, a base layer, and a subgrade. The surface layer is typically composed of cement concrete slabs and serves as the uppermost part of the pavement. It directly bears the loads from vehicles and environmental influences. The material is ordinary or high-performance concrete, mainly composed of aggregates, cement, and water. The thickness generally ranges from 18 cm to 30 cm. This layer has high strength and stiffness, serving as the primary load-bearing layer. The base layer lies beneath the surface layer. It helps distribute loads and improves the overall structural stability. Common materials include cement-stabilized crushed stone, graded gravel, or dry lean concrete. The aggregates in this layer are coarser, and the structure is relatively loose but of moderate strength. The thickness typically ranges from 15 cm to 25 cm. The bottom layer is the subgrade, which consists of natural soil and supports the entire pavement structure.

This study focuses on the debonding between the surface and base layers. To investigate this, a 2D model was developed as shown in Figure 2. Both the surface and base layers were set to a thickness of 20 cm. In this chapter, the concrete pavement is modeled as a three-phase composite consisting of aggregates, ITZ, and mortar. Based on actual 2D aggregate images, the aggregate structures in the surface and base layers are constructed using a polygonal approximation method [47]. This method offers several advantages. First, the aggregate shapes can be flexibly controlled by adjusting polygonal vertices, which provides a closer representation of the irregular geometries of real aggregates compared to circular or elliptical simplifications. Second, the approach is highly parameterized, allowing aggregate size, orientation, and distribution to be statistically defined, which enhances the randomness and representativeness of the model. Third, when combined with actual 2D aggregate images, it achieves high modeling accuracy by reproducing the spatial heterogeneity of concrete microstructure. These features make the polygonal approximation particularly suitable for defining the spatial distribution of dielectric constants in FDTD-based GPR simulations [47].

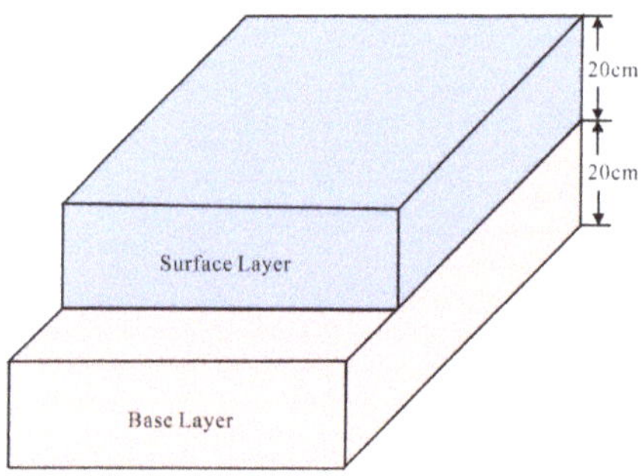

Figure 2. Schematic composition of pavement structure.

3.1. Simulation of Concrete Aggregates

To better represent the internal structure of real concrete, this study identifies the 2D contours of aggregates based on images of actual concrete slices. Figure 3 presents a representative image showing the outlines of aggregates, marked in red. As seen in the figure, the distribution of aggregates is random, and their geometries are highly irregular. Most aggregates can be approximated as irregular closed polygons, commonly with 4 to 6 sides. Such geometric characteristics are typical of natural gravel and crushed stone aggregates. These shapes are neither circular nor elliptical, and often feature sharp angles and jagged boundaries.

To accurately simulate the spatial distribution and geometric features of coarse aggregates in concrete, a large number of irregularly shaped aggregates must be generated. Traditional 2D random aggregate generation methods often rely on distance-based exclusion criteria to avoid overlapping between new and existing aggregates. However, as the number of aggregates increases, these methods suffer from low efficiency and poor convergence. To overcome these limitations, this study introduces the Fast Grid Region Division

Method (FGRDM) [47]. This method significantly improves aggregate generation efficiency and is particularly suitable for modeling high-volume-fraction concrete structures. During the modeling process, different size distributions are assigned to aggregates in the surface and base layers. Aggregates in the surface layer are smaller, while those in the base layer are relatively larger. The shapes are mainly modeled as quadrilaterals, pentagons, and hexagons. Randomization parameters include the center position, rotation angle, scaling factor, and boundary perturbation. These parameters simulate realistic particle size distributions and spatial arrangements. The resulting 2D aggregate models preserve key geometric features from actual images, improve spatial accuracy in electromagnetic simulations, and enhance structural realism. This provides a reliable foundation for radar reflection analysis and damage identification.

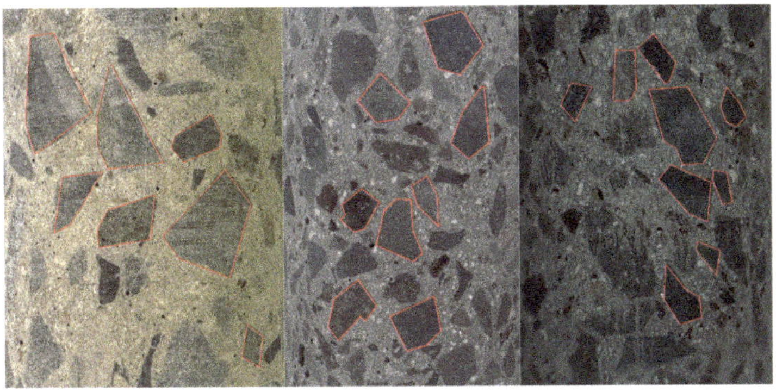

Figure 3. 2D image of real concrete aggregates.

The core idea of FGRDM is to divide the unoccupied domain into grids. Each grid is assigned a value d, representing the maximum allowable aggregate size (equivalent to the radius of the largest inscribed circle). For a candidate aggregate with radius H, one grid cell is randomly selected as the aggregate center from all grids where $d \geq H$. Once a new aggregate is generated, the occupied region is excluded from subsequent calculations. This effectively avoids overlapping. As more aggregates are generated, the available area decreases, but the generation speed increases due to the reduced number of candidate regions, thereby improving the overall modeling efficiency. This method supports both regular (e.g., ellipses) and irregular (e.g., polygons with perturbed edges) aggregate generation. It can also be extended to construct ITZ between aggregates and mortar. Compared with traditional methods, this approach ensures non-overlapping placement, accelerates generation, and improves model controllability. It provides an efficient and reliable geometric basis for electromagnetic modeling and FDTD-based ground-penetrating radar simulations of concrete. Figure 4 shows simulated aggregates with different shapes. The simulated results closely resemble the actual aggregate geometries shown in Figure 3.

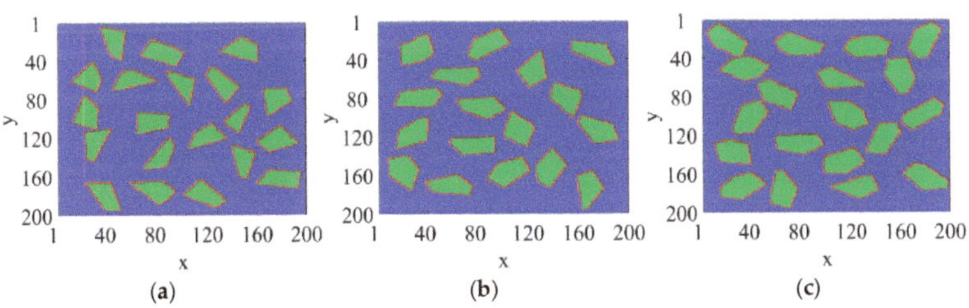

Figure 4. Simulated 2D aggregates. (**a**) Quadrilateral aggregate; (**b**) Pentagonal aggregate; (**c**) Hexagonal aggregate.

3.2. Concrete Pavement Structure

During the modeling process, the aggregate models generated in the previous section are embedded into the surface and base layers, respectively. The distinction between the surface and base layers is made primarily due to significant differences in material composition, aggregate size, and structural function. The surface layer typically uses finer aggregates with dense gradation to improve surface smoothness and wear resistance, while also bearing vehicle loads and environmental impacts. In contrast, the base layer often uses coarser aggregates to ensure structural strength and load-bearing capacity. These differences in particle size and gradation directly affect the porosity, moisture content distribution, and electromagnetic wave propagation characteristics of the concrete, leading to variations in electromagnetic parameters such as dielectric constant and conductivity between the surface and base layers. Therefore, to more accurately reflect the pavement's electromagnetic response, a more refined representation of both the surface and base layers is employed in the modeling process [52,53]. Each region is assigned corresponding electromagnetic parameters based on its material characteristics.

To reflect realistic engineering conditions, aggregate placement adheres to the mix design requirements for concrete. Different particle size distributions are used for the surface and base layers. In the surface layer, fine aggregates dominate. According to the mix design, aggregates sized 0–5 mm and 5–10 mm are added at 600 kg/m^3 and 141 kg/m^3, respectively. Coarser aggregates of 10–20 mm (515 kg/m^3) and 20–25 mm (752 kg/m^3) are also included (Table 1). The admixture is PCA-I type high-efficiency water reducer. This combination enhances surface smoothness and concrete compactness, as shown in Figure 5. In contrast, the base layer utilizes larger aggregates to enhance its mechanical performance and load-bearing capacity. Aggregates sized 20–30 mm and 10–20 mm account for 18% and 40% of the total, respectively. Smaller aggregates, ranging from 5–10 mm (15%) to 0–5 mm (27%), are also included (Table 2). Figure 6 illustrates the resulting base layer structure. Based on the aggregate gradation in each structural layer, corresponding geometric aggregate models are generated and embedded into the surface and base layer regions. This approach enables a more realistic representation of the material composition in each layer.

Table 1. Mix design of the surface layer.

Aggregate 0–5 mm (kg/m^3)	Aggregate 5–10 mm (kg/m^3)	Aggregate 10–20 mm (kg/m^3)	Aggregate 20–25 mm (kg/m^3)	Cement (kg/m^3)	Admixture (kg/m^3)	Water (kg/m^3)	*w/c* ratio	Sum (kg/m^3)
600	141	515	752	403	6	105	0.52	2522

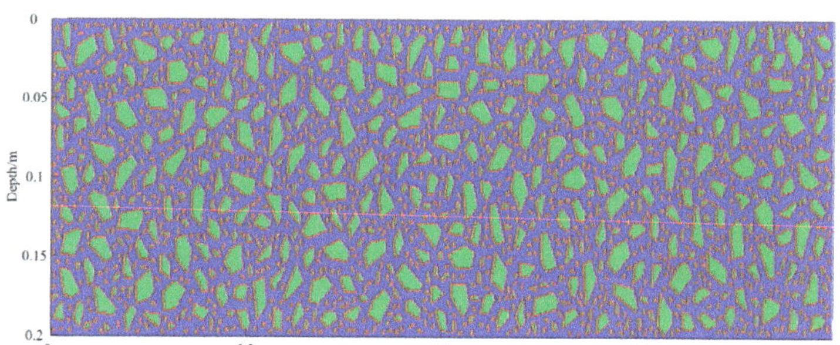

Figure 5. Simulated two-dimensional section of the surface layer block of concrete pavement. Green, red, and blue represent aggregates, ITZ, and mortar, respectively.

Table 2. Mix design of the cement-stabilized base layer.

Material Composition	Aggregate Size	Aggregate 20–30 mm	Aggregate 10–20 mm	Aggregate 5–10 mm	Aggregate 0–5 mm
	Proportion (%)	18	40	15	27
Cement Content (%)			4.5		

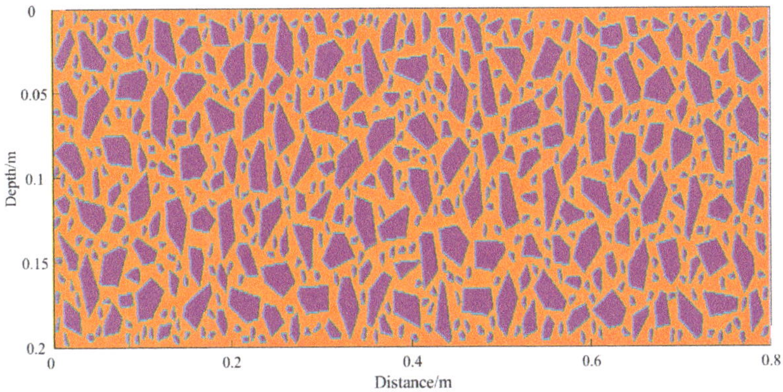

Figure 6. Simulated two-dimensional section of the base layer block of concrete pavement. Purple, cyan, and orange represent aggregates, ITZ, and mortar, respectively.

To ensure simulation accuracy and numerical stability, the model uses a uniform grid, with each cell assigned specific material properties. The dielectric constants for aggregates, cement mortar, and potential ITZ are defined separately (Table 3), resulting in a spatially heterogeneous dielectric distribution. Perfectly matched layer (PML) boundary conditions are applied at the model boundaries to eliminate artificial reflections during radar wave propagation. The core objective of this structural model is to create a computational domain that captures both the realistic geometry and dielectric heterogeneity of concrete. This provides a physically accurate foundation for analyzing the influence of interface conditions between the surface and base layers, such as debonding, on the propagation of electromagnetic waves.

Table 3. Numerical simulation models structural parameters for cement concrete pavement.

Layer	Thickness (m)	Material	Relative Dielectric Constant
Surface	0.2	Aggregate	5
		Mortar	8
		ITZ	7
Base	0.2	Aggregate	5
		Mortar	11
		ITZ	10

4. Electromagnetic Field Simulation of Different Concrete Pavement Models

To investigate the propagation behavior of GPR electromagnetic waves in concrete subgrades with irregular aggregate structures, a series of numerical models was developed using the FDTD method [46,47]. The simulation domain consists of 450 $\times$ 800 grid points, with a spatial resolution of 0.001 m. The time window is set to 15 ns. Perfectly matched layer (PML) boundary conditions are applied to eliminate artificial reflections. A 0.05 m-thick air layer is placed above the model. The source is a Ricker wavelet with a central frequency of 1 GHz and a peak amplitude of 1, polarized in the vertical (z) direction. The transmitting and receiving antennas move synchronously along the horizontal axis with a 2 mm step, generating a typical B-scan image. In B-scan images, the horizontal axis represents the movement of the antenna along the survey line, and the vertical axis corresponds to the two-way travel time of radar signals, which can be converted to depth. The amplitude of the reflected signal is shown in grayscale or pseudo-color. Subsurface anomalies such as interfaces, voids, or inclusions cause abrupt changes in electromagnetic properties, resulting in strip-like or hyperbolic reflections in the B-scan image. These features provide essential clues for target identification and defect analysis.

A total of 11 representative concrete subgrade models were constructed to simulate GPR responses under various structural configurations and voiding conditions. Model 1 assumes a homogeneous concrete structure in both the surface and base layers (Figure 7a). It serves as the reference case for analyzing wave propagation, attenuation, and reflection under ideal, uniform conditions. Model 2 introduces a void in the base layer, maintaining material homogeneity (Figure 8a). This configuration helps isolate the effects of voiding on reflection strength and time delay. Model 3 incorporates realistic concrete heterogeneity by including coarse aggregates, mortar, and an ITZ, forming a three-phase medium (Figure 9a). This model simulates scattering, velocity changes, and energy loss induced by internal heterogeneities. It also aids in distinguishing false reflections caused by material non-uniformity, improving the accuracy of void detection. Models 4 and 5 introduce voids with thicknesses of 5 mm and 10 mm, respectively, to examine how thin-layer voids affect reflection amplitude and detectability (Figure 10). Models 6 and 7 are designed to assess the role of void continuity (Figures 11a and 12a). Model 6 simulates a continuous strip-shaped void, highlighting energy superposition and interference from aligned reflections. Model 7 represents irregular or isolated voids, capturing scattering and diffraction effects that are typical of localized defects. Models 8, 9, and 10 simulate voids with thicknesses of 5 mm, 10 mm, and 15 mm, respectively (Figures 13a, 14a and 15). These configurations aim to evaluate the detectability and resolution limits of a 1 GHz radar system under heterogeneous conditions, especially for ultra-thin voids. Finally, Model 11 introduces water as a high-permittivity filler within the void to simulate moisture accumulation (Figure 16a). This scenario reflects strong signal responses and multiple echoes, offering insight into water-induced anomalies such as leakage or structural deterioration.

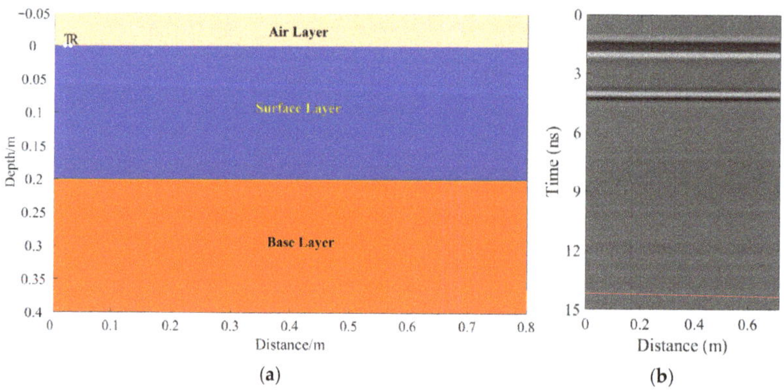

Figure 7. Model one: The surface and base layers are assumed to be homogeneous concrete. (**a**) Homogeneous concrete model; (**b**) B-scan image.

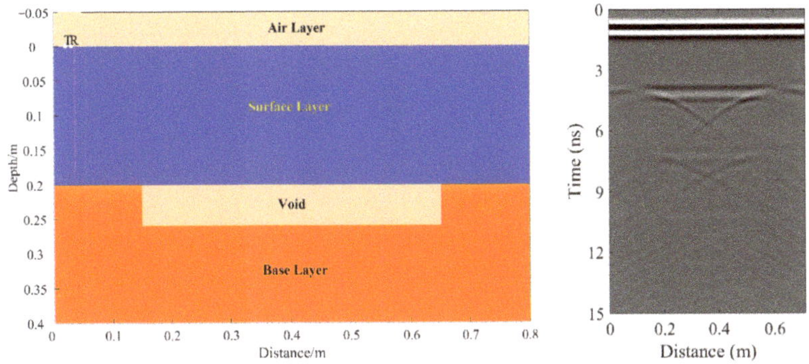

(a) (b)

Figure 8. Model two: The surface and base layers are homogeneous concrete with a void in the base layer. (**a**) Homogeneous concrete model with an embedded void; (**b**) B-scan image.

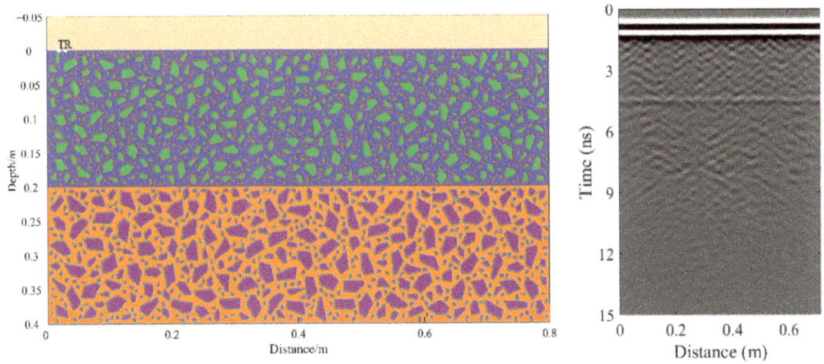

(a) (b)

Figure 9. Model three: Concrete composed of aggregates, mortar, and ITZ. (**a**) Three-phase concrete model; (**b**) B-scan image.

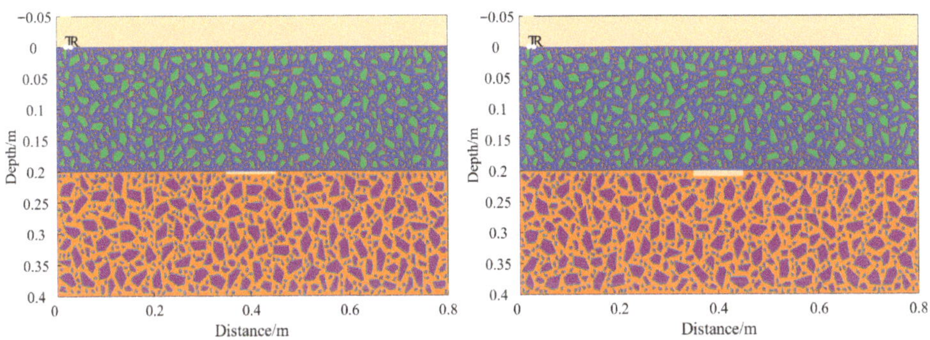

(a) (b)

Figure 10. Three-phase concrete models with different void thicknesses. (**a**) Model four: Void size 100 × 5 mm; (**b**) Model five: Void size 100 × 10 mm.

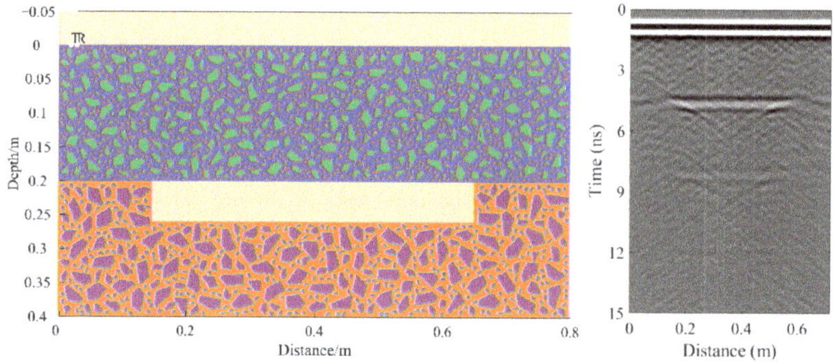

(a) (b)

Figure 11. Model six: Three-phase concrete model with continuous voids. (**a**) Continuous void model in three-phase concrete; (**b**) B-scan image.

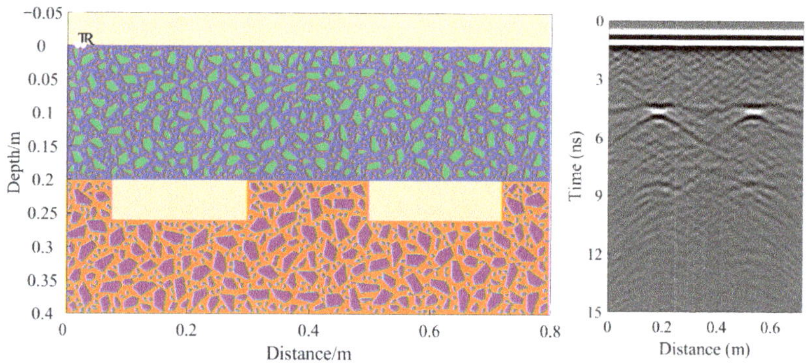

(a) (b)

Figure 12. Model seven: Three-phase concrete model with discontinuous voids. (**a**) Discontinuous void model in three-phase concrete; (**b**) B-scan image.

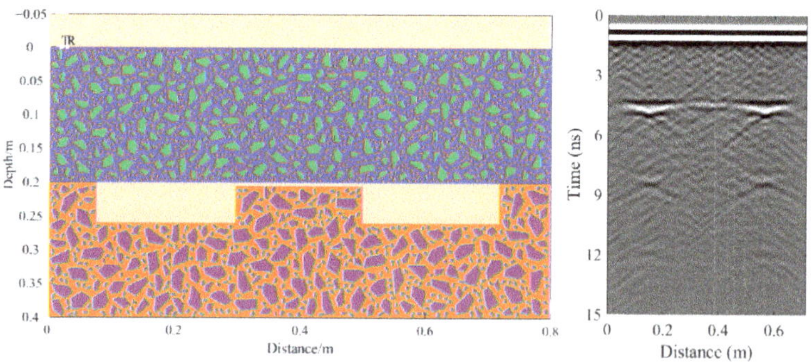

(a) (b)

Figure 13. Model eight: Three-phase model with thin voids of varying thickness. (**a**) Three-phase concrete model with 5 mm thin void; (**b**) B-scan image.

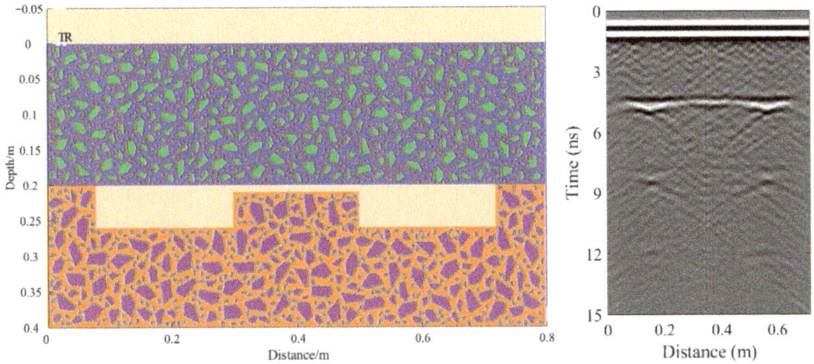

(a) (b)

Figure 14. Model nine: Three-phase model with thin voids of varying thickness. (**a**) Three-phase concrete model with 10 mm thin void; (**b**) B-scan image.

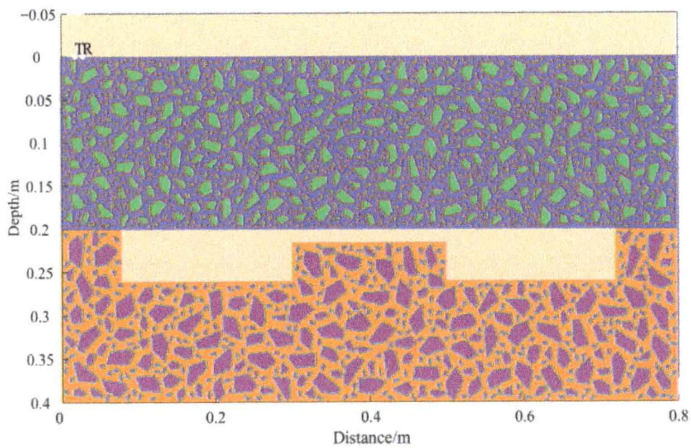

Figure 15. Model ten: Three-phase concrete model with 15 mm thin void.

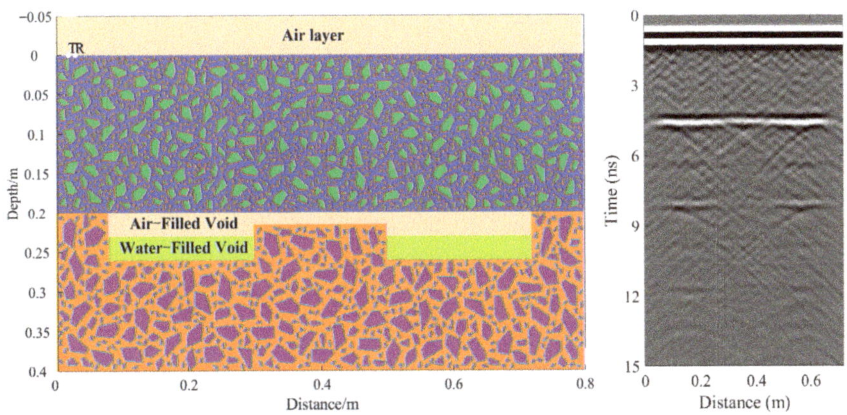

(a) (b)

Figure 16. Model eleven: Water-filled three-phase concrete model. (**a**) Three-phase concrete model with a 15 mm void filled with a 30 mm water layer; (**b**) B-scan image.

By comparing Figures 7–9, it is evident that the presence of aggregates can induce pseudo-anomalous hyperbolic reflections. In Figure 7, prominent hyperbolic waveforms appear at both ends of the void region. Furthermore, small-scale interlayer defects, such as moisture accumulation or microcracks, may also enhance radar echo intensity. However, such responses do not necessarily indicate the presence of actual voids. It is important to note that enhanced radar amplitude at interlayer interfaces does not always signify voiding. The observed increase may result from constructive interference or reflection amplification at material boundaries with differing electromagnetic properties. Therefore, void detection should be based on a comprehensive analysis that includes the shape of hyperbolic reflections, their spatial continuity, and frequency-dependent characteristics. Figure 17 presents the B-scan results corresponding to the two models in Figure 10. The left and right sections represent void thicknesses of 5 mm and 10 mm, respectively, in three-phase concrete models. Although the voids have the same horizontal extent of 100 mm, the 5 mm void produces a weaker reflection due to aggregate-induced scattering, and its hyperbolic reflection feature is not clearly visible. In contrast, the 10 mm void shows a significantly stronger echo. This suggests that aggregates can reduce radar sensitivity to thin-layer voids, potentially leading to the misidentification or omission of shallow defects during field detection. In the continuous void model shown in Figure 11, two distinct hyperbolic reflections appear in the B-scan image. These reflections are primarily associated with the boundary effects on the left and right sides of the upper void surface, which is consistent with the lateral effect of the void surface. In the discontinuous void model illustrated in Figure 12, each void has two lateral boundaries that may give rise to hyperbolic reflections, theoretically resulting in four hyperbolas. However, the simulation results reveal only two prominent hyperbolic reflections. This discrepancy can be attributed to the limitation of lateral resolution, which prevents the radar from fully separating reflections from adjacent boundaries. Therefore, we infer that the two hyperbolas in Figure 11 indeed correspond to the lateral effects of the void boundaries, while the results in Figure 12 further highlight the significant influence of limited lateral resolution on the visibility of hyperbolic reflections. This finding also underscores the necessity of cautious interpretation of boundary features in small-scale or closely spaced defects during practical engineering inspections. The analysis of Figures 12 and 13 reveals that a thin void layer located between two adjacent voids can cause superimposed reflections. This effect results in strong horizontal echoes in the B-scan, which obscure the original hyperbolic features and reduce their visibility. The comparison between Figures 13 and 14 shows that the 10 mm void produces stronger reflections than the 5 mm void. Due to aggregate scattering and signal attenuation, thin-layer voids can be easily misinterpreted as natural interlayer reflections in practice, making them difficult to identify accurately. Figure 15 shows the B-scan of a three-phase concrete model with a 15 mm void. The reflected signal from the void is significantly stronger. Figure 18 displays the B-scan results for models nine and ten. Both the 10 mm and 15 mm voids exhibit clearly identifiable thin-layer reflections. The reflection associated with the 10 mm void also presents local phase discontinuity, suggesting interface perturbations. Figure 16 corresponds to model eleven, which includes a water-filled void. Compared to model ten, the electromagnetic impedance contrast is more pronounced due to the high dielectric constant of water. As a result, the reflection amplitude is significantly enhanced, and multiple reflections are clearly observed. These results confirm that water-filled voids exhibit stronger electromagnetic responses, supporting their detectability in practical applications.

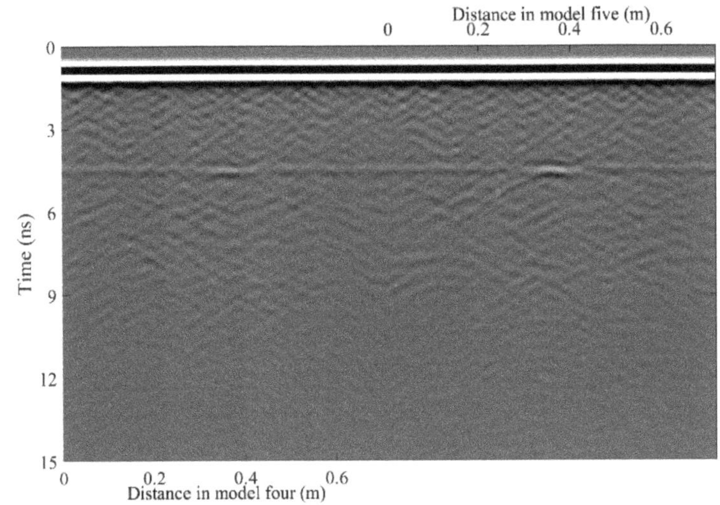

Figure 17. B-scan images of model four and model five.

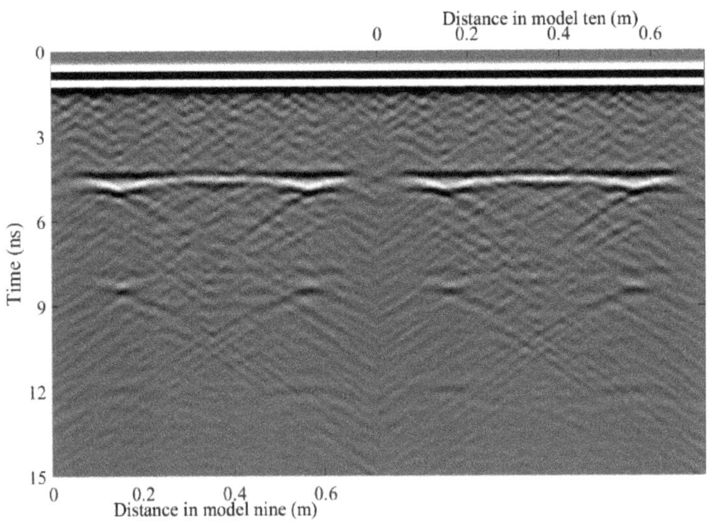

Figure 18. B-scan images of model nine and model ten.

5. Validation with Field Data

To assess the effectiveness and practical applicability of the simulation results, field tests were conducted. Two representative GPR systems were selected for joint evaluation. The first system is the GeoScope™ MK IV 3D GPR (3D-Radar A/S, Trondheim, Norway), developed by 3D-Radar, equipped with a DXG1820 ground-coupled antenna array (Figure 19a). The second is the CO1760 dual-frequency 2D GPR system from ImpulseRadar (Figure 19b). The two systems offer complementary technical advantages. GeoScope™ MK IV enables rapid, large-area scanning and structural imaging. CO1760, in contrast, provides high-resolution detection. Their combined use facilitates multi-scale and multi-depth characterization of subsurface features and anomalies.

(a) (b)

Figure 19. Radar systems used in field testing. (**a**) 3D-Radar GeoScope™ MK IV system; (**b**) ImpulseRadar CO1760 dual-frequency system.

GeoScope™ MK IV employs stepped-frequency continuous wave (SFCW) technology. It transmits a wide range of frequencies (200 MHz to 3 GHz) via a digital signal source. The DXG1820 antenna array includes 21 pairs of electronically scanned bow-tie antennas. With a channel spacing of 75 mm and a total scan width of 1.5 m, the system achieves high-density data acquisition in a single scan (Table 4). It offers both shallow, high-resolution imaging and deep-penetration capabilities. This system has demonstrated excellent performance in detecting road layers and bridge deck defects, providing continuous and clear 3D radar images that facilitate the rapid identification of subsurface variations.

Table 4. Technical specifications of the XG1820 antenna.

Antenna Model	Width (mm)	Frequency Range (MHz)	Number of Channels	Channel Spacing (mm)	Effective Scan Width (mm)
DXG1820	1800	800	20	75	1500

The CO1760 system integrates two ground-coupled antennas with central frequencies of 170 MHz and 600 MHz. It operates using continuous waveform (CWF) technology, delivering high time resolution and signal-to-noise ratio. The low-frequency channel ensures sufficient penetration depth for deep structures, while the high-frequency channel offers finer resolution for detecting shallow features such as voids and layer separations. It is particularly suitable for non-destructive testing of road structures, underground utilities, and municipal infrastructure.

In this study, GeoScope™ MK IV was first used for the rapid scanning of the test area to establish an overall subsurface profile and to locate suspicious anomalies preliminarily. Detailed follow-up surveys were then performed using the CO1760 system to enhance resolution and analyze anomaly characteristics. The combined interpretation of dual-frequency data significantly improved anomaly recognition. It was particularly effective for identifying delaminations, voids, and pseudo-reflections.

Special attention was paid to strong scattering signals that appeared similar to voids in the radargrams. Further analysis revealed that these reflections were often attributed to coarse aggregate distributions, discontinuities in the interfacial transition zone, or dielectric heterogeneity within the concrete. These are classified as pseudo-anomalies rather than true structural defects. Without cross-verification using multi-frequency responses, they could easily be misinterpreted. Therefore, systematic interpretation of waveform characteristics, spatial patterns, and frequency-dependent responses is essential. This approach helps to eliminate misleading signals and improve the accuracy and reliability of subsurface defect diagnosis.

To visualize the lateral distribution of subsurface features at different depths, C-scan images were generated using multi-frequency GPR data. Compared to traditional B-scans (vertical sections), C-scans are more suitable for capturing the horizontal extent and pattern of shallow anomalies. This enables improved detection of defects, such as voids and delaminations, thereby enhancing lateral diagnostic accuracy in concrete structures.

Figure 20 shows on-site photographs of the 3D and 2D radar systems during field surveys. The GeoScope™ MK IV system, mounted on a vehicle, was employed for rapid large-area scanning, especially suited for urban roads. Due to heavy traffic, data collection was scheduled at night. The survey vehicle operated at a controlled speed of approximately 20 km/h to minimize interference with regular traffic.

Figure 20. On-site deployment of 3D and 2D radar systems.

During the initial survey, an area with abnormal reflections was identified (Figure 21). The C-scan from the 3D GPR (Figure 21b) revealed continuous and strong reflections across all antenna channels. This anomaly exhibited significant lateral continuity and high reflection intensity, suggesting the potential presence of structural damage or a void.

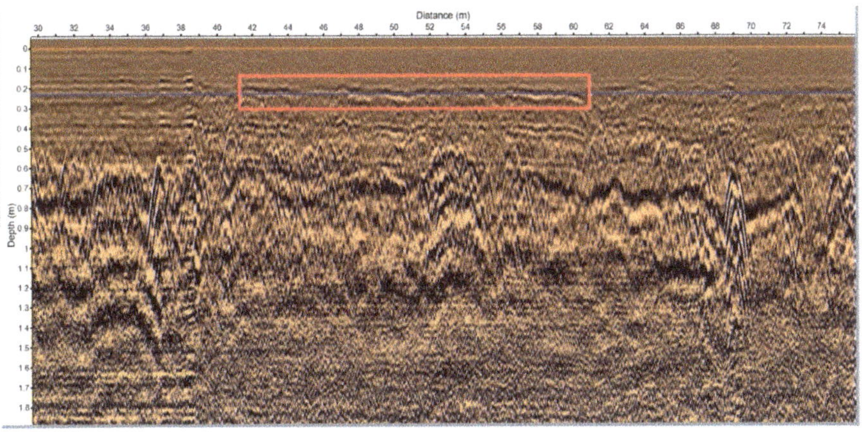

(a)

Figure 21. *Cont.*

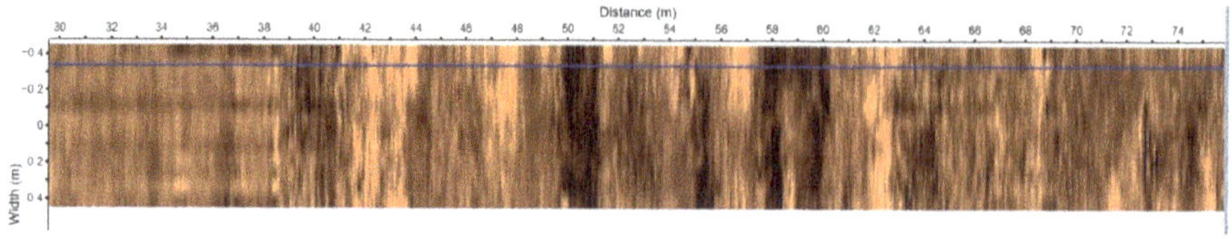

(**b**)

Figure 21. 3D GPR survey of an in-service concrete pavement. (**a**) B-scan; (**b**) C-scan.

To further examine this area, the ImpulseRadar CO1760 system was used for high-resolution resurvey. Figure 22 shows the B-scan image from the 2D radar, which captured strong reflections with waveform characteristics consistent with those observed in Figure 21a, confirming the presence of a structural anomaly. The 2D GPR image used in this study was obtained using a 600 MHz antenna, which is suitable for high-resolution imaging of shallow subsurface structures. The data were processed using Reflexw software (version 7.0), a widely used tool for GPR data analysis. In the data processing stage, a 1D filter was applied to enhance the signal quality. Specifically, the subtract-DC-shift filter was used to remove low-frequency noise and to shift the DC component, ensuring that the data reflected more accurate and meaningful subsurface features. This filtering technique is particularly effective for removing baseline drift and enhancing the clarity of reflections from subsurface structures. In both Figures 21a and 22, hyperbolic reflections at the void boundaries are clearly observed, showing strong consistency with the simulation results. This agreement not only demonstrates the capability of the simulations to reproduce real-world conditions but also highlights the effectiveness of ground-penetrating radar in identifying void boundaries, providing solid support for defect detection in engineering practice. A follow-up drilling investigation was performed to validate the radar results. As shown in Figure 23, a void was confirmed at the identified location. The void was located approximately 0.22–0.31 m below the road surface, with an estimated lateral extent of 2 m × 22 m. It was found near a maintenance cover and close to a joint between concrete slabs. Minor surface cracking was observed in the area. The anomaly likely resulted from water infiltration through surface cracks, which caused the progressive erosion of the backfill under repeated heavy traffic loads, eventually forming a cavity above the pipeline.

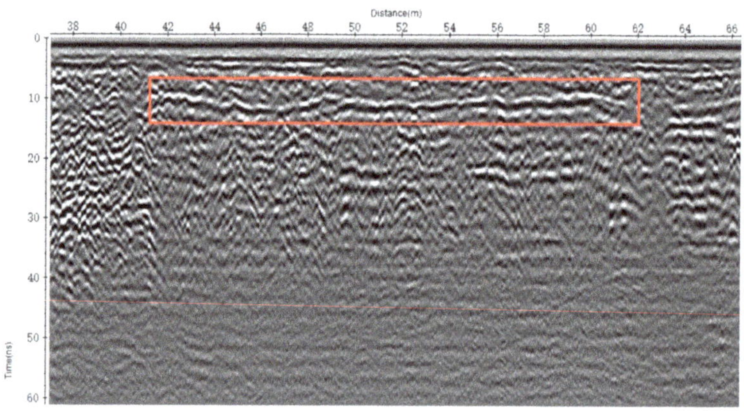

Figure 22. 2D GPR survey of the same concrete pavement (B-scan).

Figure 23. Field verification and internal view of the detected void.

During the propagation of high-frequency electromagnetic waves, the triphasic composite structure of concrete induces strong scattering, diffraction, and interference. These effects result in significant random noise and distortions in the radar profiles, which complicate the identification of actual structural interfaces. Numerous hyperbolic reflections, frequently observed in the radargrams, are typically pseudo-anomalies. These are caused not by structural defects but by localized scattering from randomly distributed coarse aggregates of varying size.

In some regions, reflections from the interface between surface and base layers appear indistinct. This is mainly due to two reasons: (1) similar dielectric constants between the two layers result in weak reflection coefficients; and (2) multiple scattering caused by coarse aggregates can mask true interlayer reflections and interfere with image interpretation.

Therefore, accurate interpretation and inversion of GPR images for concrete pavements must take into account the geometry, spatial distribution, and electromagnetic properties of coarse aggregates. Introducing a triphasic medium model allows for a more realistic simulation of radar wave propagation and energy distribution within composite materials. This enables clear differentiation between structural reflections (e.g., interfaces or voids) and material-related scattering (e.g., aggregates). By properly addressing aggregate-induced signal interference, the accuracy and physical consistency of data interpretation can be significantly improved. This provides a solid theoretical and technical basis for precise detection and quantitative assessment of pavement defects.

6. Conclusions

This study developed a triphasic concrete pavement model using the FDTD method to simulate electromagnetic responses under various structural and voiding conditions. The main findings are as follows:

(1) The heterogeneous distribution of coarse aggregates induces strong scattering during wave propagation, generating pseudo-hyperbolic echoes in B-scan images and complicating GPR interpretation with an increased risk of false positives.

(2) Aggregate-induced scattering limits the minimum detectable thickness of voids. At 1 GHz, the detection threshold is about 10 mm; thinner voids produce weak reflections that are easily masked by noise.

(3) Void geometry and continuity strongly affect radar imaging. Continuous elongated voids appear as stable planar reflections, whereas discrete small-scale voids generate localized hyperbolic patterns.

In summary, this work integrates high-fidelity triphasic modeling with multi-frequency field validation to clarify the electromagnetic response of pavement voids. The proposed model captures the influence of concrete heterogeneity on radar propagation, providing a robust basis for distinguishing structural reflections from aggregate-induced scattering. These results enhance the accuracy of void detection, reduce interpretation errors, and support the broader application of GPR in non-destructive pavement evaluation and intelligent diagnosis of defects in complex composite systems.

Nevertheless, some limitations should be noted. The model does not fully account for factors such as material moisture variability or large-scale field heterogeneity, which may influence radar responses in practical applications. Future work will aim to refine the model by incorporating these factors and validating its applicability under more diverse conditions.

Author Contributions: Conceptualization, S.W. and H.Z.; methodology, S.W. and H.Z.; software, S.W.; validation, H.Z.; formal analysis, S.W.; investigation, J.F. and H.Z.; resources, J.F. and W.H.; data curation, H.Z.; writing—original draft preparation, S.W. and H.Z.; writing—review and editing, S.W.; visualization, H.Z.; supervision, J.F. and W.H.; project administration, S.W.; funding acquisition, S.W. All authors have read and agreed to the published version of the manuscript.

Funding: This research was funded by the project ZR2022QD056 supported by Shandong Provincial Natural Science Foundation, and Qingdao Postdoctoral Science Foundation, grant number QDBSH20240101037.

Institutional Review Board Statement: Not applicable.

Informed Consent Statement: Not applicable.

Data Availability Statement: Data will be made available on request.

Conflicts of Interest: The authors declare no conflicts of interest.

References

1. Rousseau, L.S.A.; Kloostra, B.; AzariJafari, H.; Saxe, S.; Gregory, J.; Hertwich, E.G. Material Stock and Embodied Greenhouse Gas Emissions of Global and Urban Road Pavement. *Environ. Sci. Technol.* **2022**, *56*, 18050–18059. [CrossRef]
2. Zhao, G.; Zheng, X.; Yuan, Z.; Zhang, L. Spatial and Temporal Characteristics of Road Networks and Urban Expansion. *Land* **2017**, *6*, 30. [CrossRef]
3. Joshaghani, A.; Shokrabadi, M. Ground Penetrating Radar (GPR) Applications in Concrete Pavements. *Int. J. Pavement Eng.* **2022**, *23*, 4504–4531. [CrossRef]
4. Nemati, K.M.; Uhlmeyer, J.S. Accelerated Construction of Urban Intersections with Portland Cement Concrete Pavement (PCCP). *Case Stud. Constr. Mater.* **2021**, *14*, e00499. [CrossRef]
5. Joshi, T.; Dave, U. Construction of Pervious Concrete Pavement Stretch, Ahmedabad, India—Case Study. *Case Stud. Constr. Mater.* **2022**, *16*, e00622. [CrossRef]
6. Siva Rama Krishna, U.; Tadi, C. Sustainable Concrete Pavements for Low Volume Roads—Scientometric Analysis of the Literature. *IOP Conf. Ser. Earth Environ. Sci.* **2022**, *982*, 012005. [CrossRef]
7. Merten, F.R.M.; Dutra, V.F.P.; Strieder, H.L.; Graeff, Â.G. Clogging and Maintenance Evaluation of Pervious Concrete Pavements with Recycled Concrete Aggregate. *Constr. Build. Mater.* **2022**, *342*, 127939. [CrossRef]
8. Qi, G.; Wang, Z.; Chen, Y.; Zhu, C.; Ren, D.; Tian, T.; Yang, F. Analysis of Instability Mechanism and Induced Cause of Urban Pavement in Xining City, China. *Adv. Mater. Sci. Eng.* **2022**, *2022*, 3365402. [CrossRef]
9. Liao, S.-M.; Cheng, C.-H.; Chen, L.-S. The Planning and Construction of a Large Underpass Crossing Urban Expressway in Shanghai: An Exemplary Solution to the Traffic Congestions at Dead End Roads. *Tunn. Undergr. Space Technol.* **2018**, *81*, 367–381. [CrossRef]
10. Zhang, J.; Lu, Y.; Yang, Z.; Zhu, X.; Zheng, T.; Liu, X.; Tian, Y.; Li, W. Recognition of Void Defects in Airport Runways Using Ground-Penetrating Radar and Shallow CNN. *Autom. Constr.* **2022**, *138*, 104260. [CrossRef]
11. Xiong, X.; Meng, A.; Lu, J.; Tan, Y.; Chen, B.; Tang, J.; Zhang, C.; Xiao, S.; Hu, J. Automatic Detection and Location of Pavement Internal Distresses from Ground Penetrating Radar Images Based on Deep Learning. *Constr. Build. Mater.* **2024**, *411*, 134483. [CrossRef]
12. Liu, X.; Wei, X.; Qin, H. Characterizing Compressive Strength of Compacted Saline Loess Subjected to Freeze–Thaw Cycling with Wave Velocity. *Bull. Eng. Geol. Environ.* **2022**, *81*, 168. [CrossRef]
13. Qu, H.; Li, M.; Chen, G.; Sneed, L.H.; Anderson, N.L. Multi-Bandwidth Wavelet Transform of Impact Echo Signals for Delamination Detection of Thin Concrete Slabs. *J. Civ. Struct. Health Monit.* **2016**, *6*, 649–664. [CrossRef]
14. Tang, H.; Long, S.; Li, T. Quantitative Evaluation of Tunnel Lining Voids by Acoustic Spectrum Analysis. *Constr. Build. Mater.* **2019**, *228*, 116762. [CrossRef]
15. Zhao, H.; Wu, D.; Zeng, M.; Tian, Y.; Ling, J. Assessment of Concrete Pavement Support Conditions Using Distributed Optical Vibration Sensing Fiber and a Neural Network. *Constr. Build. Mater.* **2019**, *216*, 214–226. [CrossRef]

16. Wu, Y.; Liu, T.; Zhi, H.; Chen, B.; Zheng, Z.; Fang, H.; Yang, K. Experimental Study on Detection of Void in Concrete Pavement Slab by FWD. *IOP Conf. Ser. Earth Environ. Sci.* **2020**, *508*, 012080. [CrossRef]
17. Ikechukwu, A.F.; Mostafa, M.M.H. Performance Assessment of Pavement Structure Using Dynamics Cone Penetrometer (DCP). *Int. J. Pavement Res. Technol.* **2020**, *13*, 466–476. [CrossRef]
18. Yuan, J.; Jiao, H.; Wu, B.; Liu, F.; Li, W.; Du, H.; Li, J. Study of Void Detection Beneath Concrete Pavement Panels through Numerical Simulation. *Buildings* **2024**, *14*, 1956. [CrossRef]
19. Thitimakorn, T.; Kampananon, N.; Jongjaiwanichkit, N.; Kupongsak, S. Subsurface Void Detection under the Road Surface Using Ground Penetrating Radar (GPR), a Case Study in the Bangkok Metropolitan Area, Thailand. *Int. J. Geo-Eng.* **2016**, *7*, 2. [CrossRef]
20. Li, H.; Zhang, J.; Ye, M.; Wang, Q.; Wei, M.; Zhang, Z.; Guo, Y.; Tian, Y.; Zhang, Y.; Wang, C.; et al. High-Accuracy Intelligent Detection of Centimeter-Level Voids in Cement Pavement via Ground-Penetrating Radar. *Measurement* **2025**, *256*, 118354. [CrossRef]
21. Xiong, X.; Tan, Y. Deep Learning-Based Detection of Tie Bars in Concrete Pavement Using Ground Penetrating Radar. *Int. J. Pavement Eng.* **2023**, *24*, 2155648. [CrossRef]
22. Daniels, D.J. Surface-Penetrating Radar. *Electron. Commun. Eng. J.* **1996**, *8*, 165–182. [CrossRef]
23. Zhihua, H.; Tao, L.; Xiaoji, S.; Cheng, C.; Guanghu, J.; Chunlin, H.; Yi, S. A Holographic Subsurface Imaging Radar with High Plane Resolution: Principle and Application. *J. Phys. Conf. Ser.* **2024**, *2887*, 012020. [CrossRef]
24. Popov, A.; Prokopovich, I.; Kopeikin, V.; Edemskii, D. Synthetic Aperture Approach to Microwave Holographic Image Improvement. In Proceedings of the International Conference Days on Diffraction, Petersburg, Russia, 26–30 May 2014; IEEE: New York, NY, USA, 2014; pp. 192–197.
25. Wang, H.; Wu, S.; Huang, L.; Fang, G.; Liu, X. Sparse Array-Based Synthetic Spectrum Imaging Technique. In Proceedings of the 2018 17th International Conference on Ground Penetrating Radar (GPR), Rapperswil, Switzerland, 18–21 June 2018; IEEE: New York, NY, USA, 2018; pp. 1–6.
26. Windsor, C.; Capineri, L.; Bechtel, T.D. Buried Object Classification Using Holographic Radar. *Insight-Non-Destr. Test. Cond. Monit.* **2012**, *54*, 331–339. [CrossRef]
27. Tosti, F.; Ferrante, C. Using Ground Penetrating Radar Methods to Investigate Reinforced Concrete Structures. *Surv. Geophys.* **2020**, *41*, 485–530. [CrossRef]
28. Almalki, M.A.; Almutairi, K.F. Inspection of Reinforced Concrete Structures Using Ground Penetrating Radar: Experimental Approach. *J. King Saud Univ. Sci.* **2024**, *36*, 103140. [CrossRef]
29. Citir, N.; Durham, S.A.; Chorzepa, M.G.; Sonny Kim, S. Investigation of Factors Affecting Distresses on Continuously Reinforced Concrete Pavement Using Ground Penetrating Radar. *J. Perform. Constr. Facil.* **2020**, *34*, 04020027. [CrossRef]
30. Cao, Q.; Al-Qadi, I.L. Development of a Numerical Model to Predict the Dielectric Properties of Heterogeneous Asphalt Concrete. *Sensors* **2021**, *21*, 2643. [CrossRef]
31. Wang, S.; Al-Qadi, I.L. Impact and Removal of Ground-Penetrating Radar Vibration on Continuous Asphalt Concrete Pavement Density Prediction. *IEEE Trans. Geosci. Remote Sens.* **2022**, *60*, 1–14. [CrossRef]
32. Cao, Q.; Al-Qadi, I.L. Effect of Moisture Content on Calculated Dielectric Properties of Asphalt Concrete Pavements from Ground-Penetrating Radar Measurements. *Remote Sens.* **2021**, *14*, 34. [CrossRef]
33. Liu, Y.; Zhang, Z.; Yuan, Y.; Zhu, Y.; Wang, K. Quantitative Evaluation of Internal Pavement Distresses Based on 3D Ground Penetrating Radar. *Balt. J. Road Bridge Eng.* **2025**, *20*, 45–69. [CrossRef]
34. Solla, M.; Lagüela, S.; González-Jorge, H.; Arias, P. Approach to Identify Cracking in Asphalt Pavement Using GPR and Infrared Thermographic Methods: Preliminary Findings. *NDT E Int.* **2014**, *62*, 55–65. [CrossRef]
35. Loizos, A.; Plati, C. Accuracy of Ground Penetrating Radar Horn-Antenna Technique for Sensing Pavement Subsurface. *IEEE Sens. J.* **2007**, *7*, 842–850. [CrossRef]
36. Fernandes, F.M.; Pais, J.C. Laboratory Observation of Cracks in Road Pavements with GPR. *Constr. Build. Mater.* **2017**, *154*, 1130–1138. [CrossRef]
37. Hoegh, K.; Khazanovich, L.; Dai, S.; Yu, T. Evaluating Asphalt Concrete Air Void Variation via GPR Antenna Array Data. *Case Stud. Nondestruct. Test. Eval.* **2015**, *3*, 27–33. [CrossRef]
38. Chen, D.-H.; Hong, F.; Zhou, W.; Ying, P. Estimating the Hotmix Asphalt Air Voids from Ground Penetrating Radar. *NDT E Int.* **2014**, *68*, 120–127. [CrossRef]
39. AL-Qadi, I.L.; Lahouar, S. Measuring Layer Thicknesses with GPR—Theory to Practice. *Constr. Build. Mater.* **2005**, *19*, 763–772. [CrossRef]
40. Plati, C.; Loizos, A. Estimation of In-Situ Density and Moisture Content in HMA Pavements Based on GPR Trace Reflection Amplitude Using Different Frequencies. *J. Appl. Geophy.* **2013**, *97*, 3–10. [CrossRef]
41. Loizos, A.; Plati, C. Accuracy of Pavement Thicknesses Estimation Using Different Ground Penetrating Radar Analysis Approaches. *NDT E Int.* **2007**, *40*, 147–157. [CrossRef]

42. Marecos, V.; Solla, M.; Fontul, S.; Antunes, V. Assessing the Pavement Subgrade by Combining Different Non-Destructive Methods. *Constr. Build. Mater.* **2017**, *135*, 76–85. [CrossRef]
43. Tosti, F.; Bianchini Ciampoli, L.; D'Amico, F.; Alani, A.M.; Benedetto, A. An Experimental-Based Model for the Assessment of the Mechanical Properties of Road Pavements Using Ground-Penetrating Radar. *Constr. Build. Mater.* **2018**, *165*, 966–974. [CrossRef]
44. Graczyk, M.; Krysiński, L.; Topczewski, Ł.; Sudyka, J. The Use of Three-Dimensional Analysis of GPR Data in Evaluation of Operational Safety of Airfield Pavements. *Transp. Res. Procedia* **2016**, *14*, 3704–3712. [CrossRef]
45. Rasol, M.A.; Pérez-Gracia, V.; Fernandes, F.M.; Pais, J.C.; Santos-Assunçao, S.; Santos, C.; Sossa, V. GPR Laboratory Tests and Numerical Models to Characterize Cracks in Cement Concrete Specimens, Exemplifying Damage in Rigid Pavement. *Measurement* **2020**, *158*, 107662. [CrossRef]
46. Wei, S.; Wang, K.; Zhang, H.; Zhang, J.; Wei, J.; Han, W.; Niu, L. A Permeability Estimation Method Based on Elliptical Pore Approximation. *Water* **2021**, *13*, 3290. [CrossRef]
47. Wei, S.; Wang, D.; Wang, X.; Zhang, H.; Cao, H.; Liu, Q. Research on the Random Generation Model of 2D Irregular Aggregates in Concrete Based on Fast Grid Region Division Method and Its Ground Penetrating Radar Characteristics. *Case Stud. Constr. Mater.* **2024**, *21*, e03932. [CrossRef]
48. Yee, K. Numerical Solution of Initial Boundary Value Problems Involving Maxwell's Equations in Isotropic Media. *IEEE Trans. Antennas Propag.* **1966**, *14*, 302–307. [CrossRef]
49. Courant, R.; Friedrichs, K.; Lewy, H. On the Partial Difference Equations of Mathematical Physics. *IBM J. Res. Dev.* **1967**, *11*, 215–234. [CrossRef]
50. Berenger, J.-P. A Perfectly Matched Layer for the Absorption of Electromagnetic Waves. *J. Comput. Phys.* **1994**, *114*, 185–200. [CrossRef]
51. Mur, G. Absorbing Boundary Conditions for the Finite-Difference Approximation of the Time-Domain Electromagnetic-Field Equations. *IEEE Trans. Electromagn. Compat.* **1981**, *23*, 377–382. [CrossRef]
52. Soutsos, M.N.; Bungey, J.H.; Millard, S.G.; Shaw, M.R.; Patterson, A. Dielectric Properties of Concrete and Their Influence on Radar Testing. *NDT E Int.* **2001**, *34*, 419–425. [CrossRef]
53. Zhong, Y.; Wang, Y.; Zhang, B. Research on Correlation between Dielectric Properties and Compressive Strength of Concrete Based on Aggregate Particle Size. *J. Build. Eng.* **2024**, *97*, 110844. [CrossRef]

sensors

MDPI

Article

Displacement Sensing Using Bimodal Resonance in Over-Coupled Inductors

Alexis Hernandez Arroyo [1,*,†], George Overton [1], Anthony J. Mulholland [2] and Robert R. Hughes [1]

[1] School of Electronic, Electrical and Mechanical Engineering, Faculty of Engineering, University of Bristol, Bristol BS8 1TR, UK; zu18921@bristol.ac.uk (G.O.); robert.hughes@bristol.ac.uk (R.R.H.)
[2] School of Engineering Mathematics and Technology, Faculty of Engineering, University of Bristol, Bristol BS8 1TR, UK; anthony.mulholland@bristol.ac.uk
* Correspondence: alexis.hernandez@bristol.ac.uk
† Current address: Ether NDE Ltd., Endeavour House, 18 Brick Knoll Park, St Albans AL1 5UG, UK.

Abstract: This paper presents the theory and key experimental findings for investigating the generation of bimodal resonance (frequency-splitting) phenomena in mutually over-coupled inductive sensors and its exploitation to evaluate relative separation and angular displacement between coils. This innovative measurement technique explores the bimodal resonant phenomena observed between two coil designs—solenoid and planar coil geometries. The proposed sensors are evaluated against first-order analytical functions and finite element models, before experimentally validating the predicted phenomenon for the different sensor configurations. The simulated and experimental results show excellent agreement, and first-order best-fit functions are employed to predict displacement variables experimentally. Co-planar separation and angular displacement are shown to be experimentally predictable to within ± 1 mm and $\pm 1^\circ$ using this approach. This study validates the first-order physics-based models employed and demonstrates the first proof of principle for using resonant phenomena in inductive array sensors for evaluating relative displacement between array elements.

Keywords: inductors; coupling coefficients; mutual inductance

1. Introduction

Precise measurement of displacement is critical in a wide range of engineering applications. These include robotic sensing and control [1] as well as for measuring both strain within rigid structures [2] and relative displacement between components [3–5]. There are a wide range of available techniques that can be employed, depending on the requirements of the application, employing different physical phenomena. These include optical and laser methods [6–8], electrical [9,10], and electromagnetic [11] principles.

A range of different electromagnetic techniques have been explored in the literature for displacement, including capacitive, inductive, and magnetic displacement mechanisms, all using electromagnetic fields to convert relative displacement into an electrical signal [11]. One of the main techniques for measuring displacement involves the use of magneto-inductive principles [12,13], which measures the electrical properties of an inductive coil as a metallic test specimen is displaced. This relies on the change in mutual inductance between the coil and eddy currents generated within the metal sample. Jiao et al. [14] explored the mutual inductance between an LC resonator and conductive material, using this mutual inductance to calculate the separation and displacement between a resonator and the material. Similar magnetic techniques have employed magnetoresistive or Hall-effect sensors [15] to quantitatively measure the magnetic field increase or decrease from a permanent or electro-magnet due to displacement. Most of the magnetic displacement devices that are commercially available employ Hall-effect sensors, and existing technologies for human–computer interfacing measure angular and lateral displacement by evaluating the magnetic field changes of permanent magnets using Hall-effect sensors [16,17].

The tracking of resonant frequencies in inductive coils has been used to measure displacement for several applications [18,19], including by Bonfitto et al. [20] who developed a radial displacement sensor for active magnetic bearing systems.

Alternative approaches to non-contact displacement measurements include the use of wireless passive LC sensors, where strongly-coupled LC resonators present a frequency-splitting phenomenon, studied by Zhang et al. [21,22] for power transfer applications. There, the maximum efficiency is typically at the resonant frequency of the resonators; however, when similar resonators are positioned in close proximity, the mutual inductance between them increases significantly, resulting in a split in the resonance frequency and a decrease in the efficiency of power transfer. However, the phenomenon of frequency-splitting is not limited solely to power transfer applications.

The frequency-splitting phenomenon has been extensively studied in microwave antenna theory, particularly in the context of analysing displacement in microwave sensors through frequency response [23,24]. The use of the frequency-splitting phenomenon is present in different radio-frequency research studies conducted by Babu and George [25,26], where they illustrated a linear and highly sensitive displacement measurement system for wireless passive LC sensors. The frequency-splitting phenomenon manifests when inductors exhibit a substantial mutual inductance, a property determined by the magnetic coupling coefficient, k, and depends on the amount of flux sharing between inductors. As such, the frequency-splitting phenomenon is highly dependent on the distance and alignment of inductors. It has therefore been effectively applied to measure displacements in such sensors [27,28] and used across a range of different applications. Among these applications is the measurement of fluid levels inside a tank, where a passive resonator coil floats at the liquid's surface and the frequency-splitting effect is measured in an external coil [27]. Frequency-splitting frequency has also been employed in medical applications [29] where an optimum frequency tracking method ensures that the inductive link is working at resonance and the voltage and efficiency are at maximum.

The use of magnetic coupling for evaluating displacement is not limited to the use of two LC resonators. To date, the evaluation of displacement in inductive resonant sensors has primarily focused on coaxial air-core coils. This paper seeks to extend the principles of resonant frequency-splitting to over-coupled inductors in planar array configurations to evaluate relative displacement (separation and angular) in adjacent coils. In this paper, we exploit the magnetic coupling enhancing effects of ferrite cores and employ simple approximations to link the frequency-splitting to the relevant displacement variables via the coupling coefficient.

The proof of principle is demonstrated in a two-coil configuration, where the evaluation of the coupling coefficient is made by monitoring the frequency spectra response in the active branch of LC sensors. The wireless and single-branch measurement allows the application of the sensors in different technologies, yet to be explored. The different coil configuration extremes demonstrated in this paper highlight the accuracy and versatility of the modelling for designing resonant displacement sensors.

2. Theory

Inductive array sensors operate on the principle of electromagnetic induction, where alternating currents generate changing magnetic fields via Ampere's law. These changing magnetic fields generated by a transmitter coil generate induced currents in neighbouring elements [30]. For closely spaced coils, this effect generates mutual coupling between them. This paper focuses on the prediction, measurement, and characterization of coupling between identical resonant inductors for the evaluation of displacement in inductive sensors. This exploits a phenomenon of bimodal resonance observed by Zhang et al. [25,31] and Hughes et al. [32] in closely packed inductors. The following subsections outline the principles behind mutual coupling and resonance in two-coil probes.

2.1. Self-Resonance Equivalent Circuit Model

Inductive sensors can be modelled as a parallel LC equivalent circuit (see Figure 1a), with the inductor, L_1, and the capacitor, C_1, representing the energy stored in the electrical field within the sensor, respectively [33]. Typically, C_1 is the lumped combination of capacitive effects between coil windings as well as any cabling or capacitive loads applied to the sensor. The resonant frequency, f_0, can be defined for a single inductor circuit as

$$\omega_0 = 2\pi f_0 = \frac{1}{\sqrt{LC}}, \tag{1}$$

where ω_0 is the angular resonant frequency. At this natural resonance, the magnitude of the impedance, $|Z|$, of the sensor is a maximum, and the voltage, V, and current, I, within the circuit are in phase (zero phase lag).

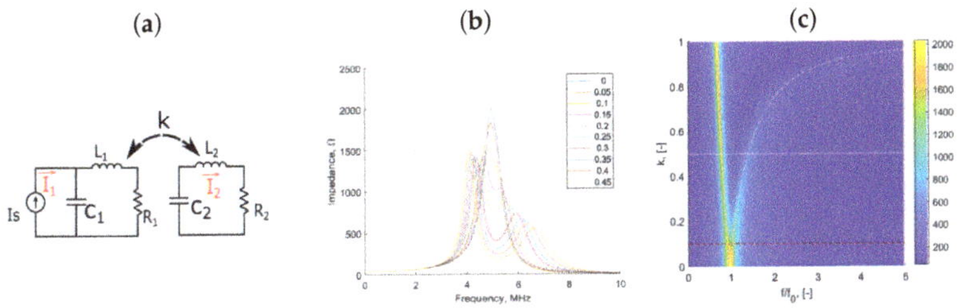

Figure 1. Equivalent circuit modelled results of a two-coil over-coupled inductor system, showing (**a**) the equivalent circuit diagram, (**b**) the circuit model's predicted bimodal resonance phenomenon exhibited in the electrical impedance magnitude spectra as a function of the coupling coefficient, k, (from Equation (8)), and (**c**) impedance magnitude heat map showing the trajectories of resonant peaks as a function of k. Circuit component values used are $R_1 = R_2 = 100\ \Omega$, $C_1 = C_2 = 1$ nF, and inductance of $L_1 = L_2 = 160\ \mu$H.

Operating inductive sensors at or near resonance have been shown by many authors to improve measurement sensitivity and maximise the mutual inductance between array elements [34–36]; however, the resonant measurement techniques in array sensors have yet to be explored in detail.

2.2. Two-Coil Mutual Resonance Model

When close to another comparable coil, the primary coil will inductively couple to the secondary coil (see Figure 1a). The secondary coil can be modelled as an inductor, L_2, in series with a resistor, R_2, and a capacitor, C_2. This coupling, parametrized by the coupling coefficient, k (see Equation (10)), will alter the effective inductance and resistance (L_1' and R_1', respectively) of the primary measurement circuit and will distort the measured impedance, Z_m', given as [34]

$$Z_m' = \frac{R_1' + i\omega L_1'}{1 + i\omega R_1' C_1 - \omega^2 L_1' C_1}. \tag{2}$$

Here, L_1' and R_1' are given by

$$L_1' = L_1\left[1 - \alpha^2\frac{L_2}{L_1}\left(1 - \frac{\omega_2^2}{\omega^2}\right)\right], \tag{3}$$

$$R_1' = R_1\left[1 + \alpha^2\frac{R_2}{R_1}\right], \tag{4}$$

where α^2 is defined as

$$\alpha^2 = \frac{\omega^2 M^2}{R_2^2 + \omega^2 L_2^2\left(1 - \frac{\omega_2^2}{\omega^2}\right)^2}, \tag{5}$$

where $M = k\sqrt{L_1 L_2}$ is the mutual inductance between coils. The resonant frequencies of the two-coil system can be derived from Equations (2)–(5) to give the resulting expression (see Appendix A):

$$\omega_\pm' = \sqrt{\frac{1}{2}\frac{\omega_1^2}{[1-k^2]}\left[1 + \frac{\omega_2^2}{\omega_1^2}\left(1 \pm \frac{\omega_1}{\omega_2}\sqrt{\frac{\omega_2^2}{\omega_1^2} + \frac{\omega_1^2}{\omega_2^2} + 4k^2 - 2}\right)\right]}, \tag{6}$$

where $\omega_n = \sqrt{1/L_n C_n}$ is the uncoupled resonant frequency of coil n. Equation (6) predicts two distinct resonant frequencies of the two-coil system. These can be determined for a given value of k by knowing the natural resonant frequency of each coil in their uncoupled ($k = 0$) state.

For the special case of identical coils where $L_1 = L_2$ and $C_1 = C_2$, or more specifically when their resonant frequencies are matched $\omega_1 = \omega_2$, Equation (6) simplifies to give the formula for the coupled resonant frequencies $\omega_\pm'$ of

$$\omega_\pm' \approx \sqrt{\frac{1}{C_1 L_1 [1 \pm k]}} = \omega_1\sqrt{\frac{1}{1 \pm k}}. \tag{7}$$

Equation (7) can be used to predict the change in the resonant frequencies as a function of k [25]. Figure 1b shows how the magnetic coupling k affects the resonant frequency and how it produces the resonant frequency-splitting phenomenon that separates the resonant peaks, shown as white dashed lines in Figure 1c, as a function of the coupling coefficient. The system can be thought of as exhibiting independent vibrational modes. Figure 1c also shows a red dotted line representing the dispersion separation threshold—the coupling coefficient of this system—above which resonant peaks can be resolved as distinct peaks. This threshold is dependent on the quality factor of the systems and, as such, lower resistance systems exhibit sharper, more easily resolvable curves at lower coupling coefficients. There is also a practical upper threshold to the coupling coefficient of a realistic inductively coupled system which is dependent on the geometry of the system and the permeability of the cores used within the coils.

From Equation (7), an expression for the coupling coefficient, k, between identical coils can be derived as a function of the measurable resonant frequencies observed in the spectra:

$$k \approx \frac{f_-^2 - f_+^2}{f_-^2 + f_+^2} \approx \frac{\gamma_\pm^2 - 1}{\gamma_\pm^2 + 1} \tag{8}$$

where $\gamma_\pm = f_-/f_+$ is the bimodal resonant frequency ratio. Equation (8) matches the generalised Cohn–Matthaei formula for coupled resonators derived by Tyurnev 2007 [37]. Comparable formulas are regularly applied in the modelling of microwave bandpass networks [38] and meta-material design [39], but the features and properties of coupled resonators are yet to be explored in inductive measurement applications. Rearranging Equation (7) gives an expression for the bimodal frequency ratio, $\gamma_{\pm}$, between split resonant frequencies, as a function of k, for an identical coil configuration:

$$\gamma_{\pm} = \frac{f_-}{f_+} \approx \sqrt{\frac{1+k}{1-k}}. \tag{9}$$

2.3. *Coupling Coefficients*

Using the centre of magnetism (CoMag) approach detailed in [40], a 2D approximation for the coupling coefficient, k, between neighbouring coils (see Figure 2) can be expressed as

$$k = \frac{\Phi_2}{\Phi_1} \approx \frac{1}{4\pi}\int_{r'_{2a}}^{r'_{2b}} \frac{1}{r'}\,dr' \approx \frac{1}{4\pi}\ln\left[\frac{r'_{2b}}{r'_{2a}}\right], \tag{10}$$

where Φ_n is the magnetic flux through coil n, r_1 and r_2 are the primary (driver) and secondary (passive) coil radii, h is coil height, a is coil separation distance, and $\theta \in [\pi/2, \pi]$ ($\theta = \pi - \phi$) is the relative angle. Variables r'_{2a} and r'_{2b} are the distances from the CoMag to the centre of the cross-section of each side of the coil (subscript a and b denoting the near and far sides of the passive coil, respectively), and can be defined as

$$r'_{2a} = r_2\Lambda\sqrt{2(1-\cos\theta)}, \tag{11}$$

$$r'_{2b} = r_2\sqrt{2[2 + 2\Lambda(1-\cos\theta) + \Lambda^2(1-\cos\theta)]}, \tag{12}$$

where

$$\Lambda = \zeta + \eta\tan\frac{\phi}{2}, \tag{13}$$

is an intermediate term relating to the geometric ratios of coil configuration, with $\zeta = a/r_2$ as the separation ratio, and $\eta = h/r_2$ is the aspect ratio of the coil. Substituting the above expressions into Equation (10) allows us to calculate k as a function of the separation, a, or relative angle θ. Three specific cases can be considered: co-planar separation ($\theta = \pi$), the angular displacement of planar coils ($h \ll r_2$), and angular displacement of solenoid coils ($h \gg r_2$).

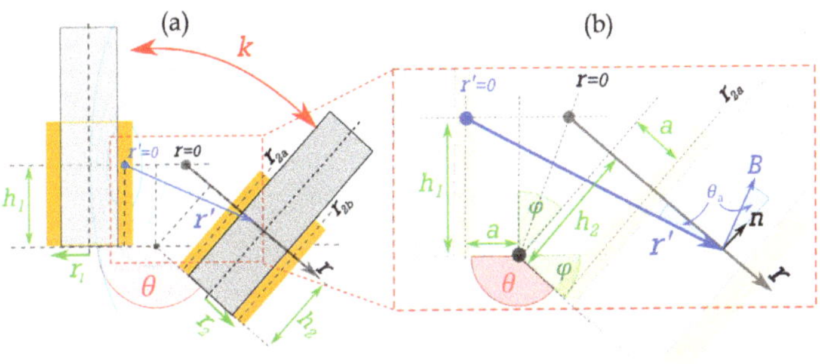

Figure 2. Diagram of coupled 2D ferrite-cored solenoid coils, where the yellow blocks represent the cross-section of the coil, and grey cores represent ferrite: (**a**) angular separation θ between the centre of magnetism, h, of each coil, and (**b**) a zoomed-in diagram of the trigonometric relationship between the distance from the line of symmetry ($r = 0$ mm) and the distance from the centre of the B-field source ($r' = 0$ mm), showing separation, a, from the centre-point ($r = 0$ mm), and angular displacement $\phi = \pi - \theta$ (from [40]).

2.3.1. Co-Planar Separation

When identical coils are co-planar ($\theta = \pi$), $\Lambda = \zeta$, and Equations (11) and (12) can be simplified to $r'_{2a} = 2a$ and $r'_{2b} = 2r_2(\zeta + 1)$. We can therefore define the coupling coefficient from Equation (10) as a function of the dimensionless separation ratio, ζ:

$$k \approx \frac{1}{4\pi} \ln\left(\frac{\zeta + 1}{\zeta}\right) \equiv \frac{1}{4\pi} \ln\left(1 + \frac{r_2}{a}\right). \tag{14}$$

Equation (14) is valid for all coil aspect ratios. Equation (14) can be rearranged to arrive at a generalised linear function of the form $y = p_1 x + p_2$:

$$e^{4\pi k} \approx p_1 \frac{1}{a} + p_2, \tag{15}$$

where p_1 and p_2 are unknown coefficients of the first-order polynomial and can be fitted to experimental data to enable displacement prediction [40].

2.3.2. Angular Displacement—Planar Coils

Consider two planar coils with a varying angle as is shown in Figure 2, with $h \ll r_2$, such that the coil aspect ratio can be considered negligible ($\eta \ll \zeta$) and the coils are planar. In this case, for an angle ϕ away from co-planar existence (i.e., $\theta < \pi$), $\Lambda \approx \zeta$, such that Equations (11) and (12) can be simplified to express the coupling coefficient, as follows [40]:

$$k \approx \frac{1}{8\pi} \ln\left[1 + \frac{2}{\zeta} + \frac{4}{\zeta^2(4 - \phi^2)}\right], \tag{16}$$

where ϕ is given in radians. Equation (16) can be simplified into a generalised linear function for the angular displacement, ϕ, for filament coils:

$$e^{8\pi k} \approx p_1 \frac{1}{(4 - \phi^2)} + p_2, \tag{17}$$

where again p_1 and p_2 are the unknown coefficients of a first-order polynomial, which can be found by fitting to experimental data [40].

2.3.3. Angular Displacement—Solenoid Coils

For the case when $a \ll h$ and $\zeta \ll \eta$, a first-order approximation can be derived for k as [40]

$$k \approx \frac{1}{4\pi} \ln\left[1 + \frac{2}{\eta\phi}\right]. \tag{18}$$

As before, a generalised linear function for ϕ can be defined as

$$\eta e^{4\pi k} \approx p_1 \frac{1}{\phi} + p_2. \tag{19}$$

Equations (14) and (18) provide simple functions with which to predict ϕ between coils when k is experimentally measurable. While the extreme approximations used to arrive at these functions mean that they may not provide accurate absolute values, their generalised functions given in Equations (17) and (19) can be used to fit experimental data [40].

3. Materials and Methods

Finite element modelling (FEM) and experimental tests were conducted to validate and characterise the bimodal resonance behaviour predicted by the circuit theory

(Section 2.2). The complete details of the modelling and experimental methods are given in the following subsections.

3.1. Sensor Designs

To test the previously defined equations, two different extremes of coils were developed, with two different studies: co-planar separation and angular displacement.

For construction simplicity, the solenoid coil was designed with a single layer of forty turns with a wire diameter of 0.56 mm, and the planar coil was a double-layer printed circuit board (PCB) with a rectangular shape (see Figure 3d). The electrical and geometrical parameters of the solenoid and planar coils are shown in Table 1. Experimental coil dimensions are given with an error of ±0.5 mm, and experimental circuit parameters are given with tolerances of ±5%.

The coils are connected to a resonant tank circuit [41], commonly used for radio frequency and also in signal filtering applications. The solenoid coils were fabricated in-house, while the planar coils were manufactured by PCBway, China.

The inductance of the coil was obtained using an impedance analyser TREWMAC TE3001 (TrewMac Systems, Adelaida, Australia). For the 2D model and 3D model, the geometry coil analysis module in COMSOL 6.1 was used to obtain the values of the inductances, and the capacitor and resistor used the experimental values obtained previously.

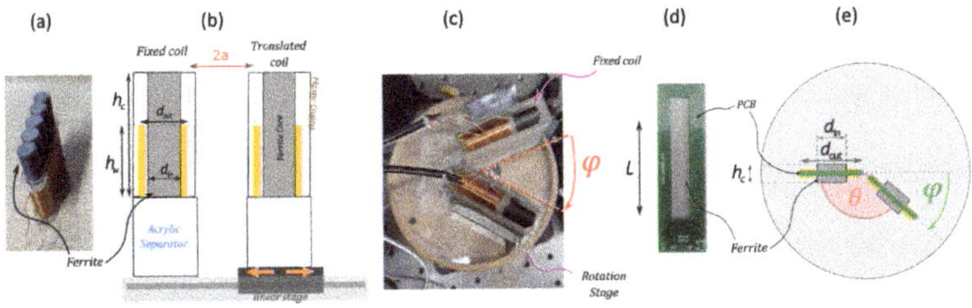

Figure 3. Coil design and experimental setup for co-planar separation and angular displacement studies. (**a**) Solenoid ferrite-cored coil top-down design, along with (**b**) cross-sectional co-planar separation (a) and (**c**) angular displacement (ϕ) testing configurations. (**d**) PCB-type (planar) coil top-down design and (**e**) angular displacement experimental configuration.

Table 1. Experimental (Exp.) and simulated sensor coil dimensions and electrical circuit parameters for solenoid (Sol.) and planar (Plan.) coils in experimental tests, as well as 2D and 3D finite element simulations.

Coil Dimensions (mm)					
Coil Type	l	h_c	h_w	d_{in}	d_{out}
Sol	43	40	25	8	11
Plan.	20	1.5	1.0	5	12
Circuit Parameters					
Coil Type	Turns	L (µH)	C (nF)	R (kΩ)	I (mA)
Sol. (2D)	40	708	0.2	0.1	10
Sol. (3D)	40	170	1	1	10
Sol. (exp.)	40	166	1	1	10
Plan. (2D)	28	7.9	0.9	0.01	1
Plan. (3D)	28	7.5	1	0.01	1
Plan. (exp.)	28	6.7	1	0.01	1

3.2. Finite Element Modelling

Parametrized finite element models (FEMs) were created in COMSOL 6.1 using the AC/DC module. While this paper is not focused on the optimization of the magnetic field in inductors, the magnetic coupling of the inductors is directly related to the generation of the bimodal phenomenon. Models were developed using magnetic field and electrical circuit studies to simulate the bimodal phenomenon. These are compared to experimental measurements of two extreme coil geometries, planar and solenoid coils.

The parameters used for the modelled bimodal sensors are shown in Table 1. Figure 4 shows an example simulated frequency spectrum and how the magnetic flux density, B, changes within an over-coupled two-coil system as a function of frequency, f. It is evident that at the first resonant peak ($f = f_+$), the magnetic flux is predominantly emitted from the driver coil (Figure 4ii), while when $f = f_-$, the flux is emitted predominantly from the passive coil (Figure 4iii).

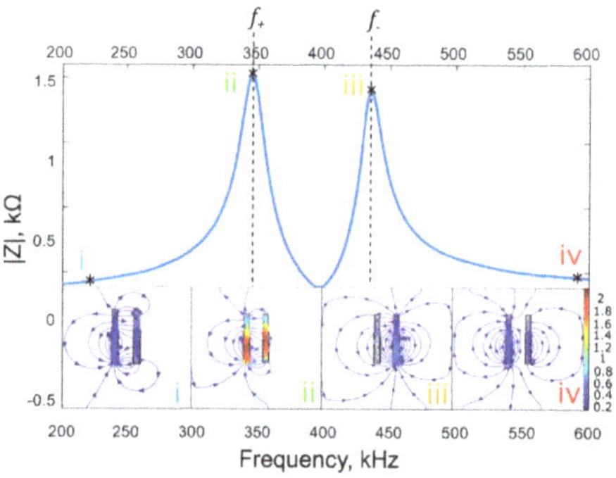

Figure 4. Finite element simulation of the magnetic flux density, B, in a bimodal resonant system in relation to the electrical impedance magnitude $|Z|$ as a function of frequency, f. The spatial distribution of the in-plane B-field is shown at (i) 225 kHz, (ii) 350 kHz (first resonant peak, f_+), (iii) 450 kHz (second resonant peak, f_-), and (iv) 590 kHz.

The magnetic flux density shown in Figure 4ii is maximum at the first resonance peak, which corresponds to a maximum B-field in both the driver and passive coil, generating maximum power transfer. This effect of maximum power transfer has already been explored by Zhang et al. [42].

Figure 4i shows how the B-field is concentrated around the driver coil (left) for $f < f_+$ of the coils. Figure 4ii shows that at $f = f_+$, B increases significantly for both coils but the spatial distribution remains comparable. Figure 4iii shows that B is concentrated in the passive coil at $f = f_-$, and Figure 4iv shows how B is distributed evenly at higher frequencies beyond the resonant peaks.

3.3. Experimental Measurements

The two sensor configurations defined in Table 1 had their bimodal impedance spectra evaluated experimentally as a function of their co-planar separation, a, and angular displacement, ϕ. The equipment and methods used are summarised below.

3.3.1. Data Acquisition

A TREWMAC TE3001 (TrewMac Systems, Australia) impedance analyser was used to measure the impedance across the driver coil. The impedance analyser took a total of 1024 points in a frequency sweep between 0.3 and 0.5 MHz for the solenoid coils and between 1.7 and 2.2 MHz for the planar coils. These data were recorded, and, using an automated peak detection process, the frequencies of the bimodal resonant peals were determined. For each angle, a frequency sweep is recorded using the impedance analyser, and the relative differences between the bimodal resonance peaks are determined. As shown in Figure 3b, the coils are contained in a plastic fixture of a wall thickness of 1.5 mm which generates an initial separation between the coils. The results of the initial impedance frequencies for solenoid and planar coils are shown in Figure 5.

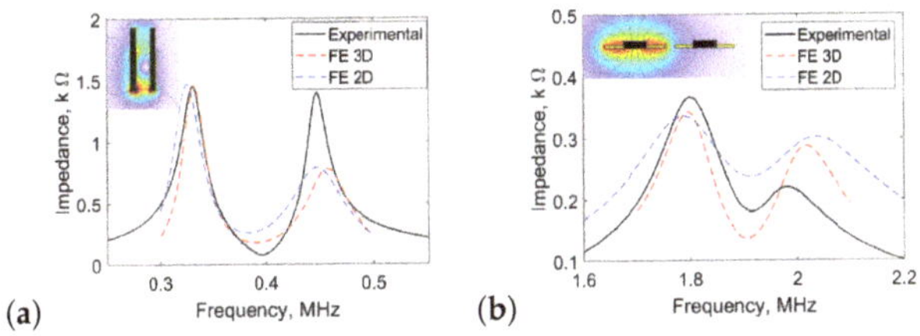

Figure 5. Comparison between simulated and experimental electrical resonant impedance spectra in an over-coupled driver coil. Graphs compare experimental results to 2D and 3D finite element modelled results (blue and red dashed curves, respectively) for (**a**) solenoid coils and (**b**) planar coils.

3.3.2. Co-Planar Separation, a

An X-LSM200A-E03 (Zaber, Vancouver Canada) linear stage was attached to the passive sensor (see Figure 3b), and moved in increments of 10 mm, within a coil centre-to-centre distance range of 15–75 mm (beyond which the coupling is too low for the resonance frequency to be observed in the resonance frequency sweep). A second measurement study was conducted with separation increments of 1 mm to provide a higher-resolution dataset used to predict the separation between the sensors.

3.3.3. Angular Displacement, ϕ

An X-RSB120AU rotational stage (Zaber, Vancouver, BC, Canada) was used to incrementally rotate a passive resonator coil relative to a fixed driver coil, with the centre of rotation at the vertex of the shared corner, as shown in Figure 3c for solenoid coils and in Figure 3e for planar coils. Two sets of measurements were conducted: Firstly ϕ, was varied between 0 and 70° in 10° increments; the resonant frequencies were recorded and a best-fit function (derived from the formula in Section 2.3) was applied to the data. The resonant frequencies were measured again, this time for ϕ in increments of 1°, and the best-fit functions (see Section 2.3) were used to invert ϕ.

4. Results and Discussion

Experimental and FE results are compared directly for each of the studies, exploring co-planar separation, a, and angular displacement, ϕ. The results show findings varying the co-planar separation, a, for solenoid coils, and ϕ for both solenoid and planar coil configurations. The results are compared to best-fit functions defined in Section 2.

The experimental impedance spectra of the bimodal sensors for the two coil geometries are compared to the FE models in Figure 5. Based on the resonant splitting phenomena, the relationship between bimodal resonant frequencies and displacements a and ϕ can be evaluated.

Equation (8) is used to calculate k from measurable resonant frequencies, f_- and f_+, and then the first-order approximations (Section 2.3) can be used to estimate displacement variables.

4.1. FE Model Validation

The experimental impedance spectrum results were compared to FE simulations in 2D and 3D, showing differences in the values of capacitance, as shown in Table 1, due to the separation of the wire in the hand-wound coils. These changes in capacitance lead to a different resonant frequency. To match the resonant frequencies, it is necessary to introduce a calibration value of capacitance for the resonators. This allows the simulated system to resonate at the same frequencies as the experiment.

For the solenoid and planar coils, the resonant frequencies for the three different results are approximately matched in frequency value. The magnitude depends on the impedance of the system; for the solenoid coil, there is a better approximation of magnitude than for the planar coil. Figure 5a shows the frequency spectrum for the solenoid coils with an initial centre-to-centre separation of 1.5 mm; this separation a represents the separation between the edges of the windings in the coils (see Figure 2).

For an initial separation of $a = 3$ mm between the solenoid windings due to the plastic case, the solenoid coils produce the initial bimodal resonance frequency spectrum shown in Figure 5a. Planar PCB coils, with an initial separation of $a = 2$ mm, produce the bimodal frequency spectrum shown in Figure 5b. When the coil decreases in size, the number of turns reduces, making it harder to match the finite element simulation with the experimental data due to the homogeneous multiturn approximation.

The difference between experimental and simulated results is predominantly due to disparities between real and simulated values for key variables—for instance, the ferrite permeability, lumped circuit component values, and other estimations that were made in the simulation. While the spectral responses simulated do not precisely map onto the experimental values, the simulations are sufficient to demonstrate the operational principle of using the resonance splitting frequencies for displacement applications.

4.2. Co-Planar Separation, *a*

The bimodal resonant frequencies f_- and f_+ were used to determine k from Equation (8) as a function of a. As a increases, k decreases, causing f_- and f_+ to converge in the frequency spectrum until they merge, as demonstrated in Figure 1b.

For the 2D and 3D FE, the bimodal frequency spectrum merges into a single frequency at separations greater than 12.5 mm and 17.5 mm, respectively, where the passive coil has minimal interaction with the driver coil. This phenomenon has been explored by Zhang et al. [21] for wireless power transfer applications.

Despite the differences in the impedance frequency spectrum for the 3D and 2D simulation (Figure 5), the resonant frequencies exhibit similar trends as a function of changes in coupling, as shown in Figure 6c.

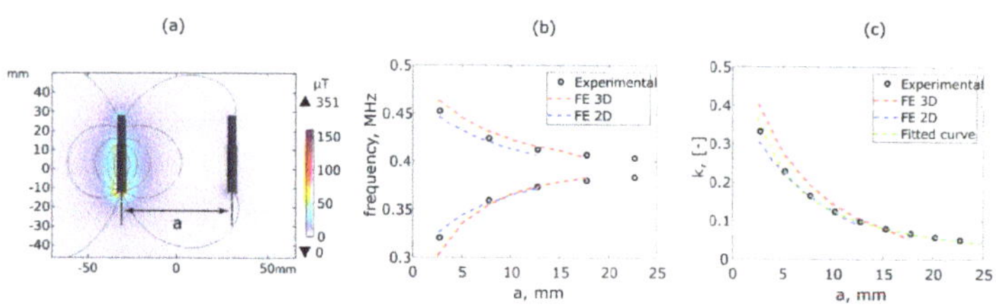

Figure 6. Bimodal resonant frequencies as a function of co-planar separation, a: (**a**) 2D finite element modelled magnetic flux density and field lines between bimodal sensor elements for a separation of $a = 60$ mm in solenoid coils. (**b**) Bimodal resonant frequencies, f_- (lower curves) and f_+ (upper curves), of the system as a function of angular displacement, comparing experimental (black circles) to 2D (blue dashed) and 3D (red dashed) finite element simulated results. (**c**) The coupling coefficient, k, calculated using measurements of f_- and f_+ from Equation (8), where the green dashed line represents the experimental best-fit function, as defined by Equation (15) in Section 2.3.

4.3. Angular Displacement, ϕ

The results in Figure 7a(ii) show how the resonant frequencies f_+ and f_- get closer together as a function of ϕ for the solenoid coils, showing comparable results to the evaluation of co-planar separation, a. In this case, k decreases as ϕ increases. Figure 7b shows f_+ and f_-, as a function of ϕ for the planar coils. In this case, k increases as ϕ increases.

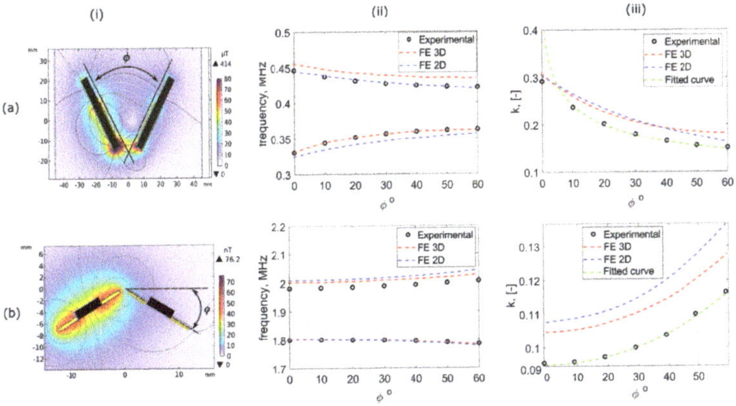

Figure 7. Bimodal resonant frequencies as a function of angular displacement, ϕ, in (**a**) solenoid and (**b**) planar coils: (**i**) 2D finite element modelled magnetic flux density and field lines between bimodal sensor elements for a separation of $\phi = 60°$ in solenoid coils. (**ii**) Bimodal resonant frequencies, f_+ (lower curves) and f_- (upper curves), of the system as a function of angular displacement, comparing experimental (black circles) to 2D (blue dashed) and 3D (red dashed) finite element simulated results. (**iii**) The coupling coefficient, k, calculated using measurements of f_- and f_+ from Equation (8), where the green dashed line represents the experimental best-fit function, as defined by Equations (19) and (17) in Section 2.3.

The k between the solenoid coils varies between 0.15 and 0.35, as shown in Figure 7a(iii). This is relatively high for non-coaxial coils, where a typical value is $k < 0.1$ [43]. These high k values are due to the large ferrite cores and the significant magnetic bridging it causes between the coils. The planar coils exhibit more modest values of k, between 0.9 and 0.14. As k between the planar sensors is relatively low compared to the solenoid coils (see Figure 7b(iii)), the matching between the FE models and experiment becomes more challenging, leading to an offset between results for k of the planar coil geometry.

The angular separation of the ferrite cores in solenoid coils increases the distance between the centres of magnetism of the cores (see Section 2.3), decreasing k. Conversely, the distance between the centres of magnetism for the planar coils reduces as ϕ increases, causing an increase in k. These trends are clearly shown in Figure 7a(iii),b(iii). Figure 7 therefore shows excellent agreement between the experimental and FE simulations, with f_+ and f_- changing as predicted in Section 2.3.

4.4. Displacement Prediction

The experimental displacement measurements shown in Figures 6 and 7 for co-planar separations, a, and angular displacements, ϕ, with increments of $\Delta a = 10$ mm and $\Delta\phi = 10°$, respectively, were used to fit the first-order functions defined in Section 2.3.

From the values of k calculated using Equation (8), linear functions of the form $y = p_1 x + p_2$ were plotted based on Equations (15), (17) and (19), as defined in Table 2 and [40]. Linear regression was then used to calculate the coefficients p_1 and p_2 shown in Table 2. Note that the factor of 4 in Equation (15) was omitted as this gave rise to a stronger linear correlation as found by [40]. This disparity is likely due to the simplicity of the assumptions employed in the derivations of the first-order functions derived in [40] (see Figure 8).

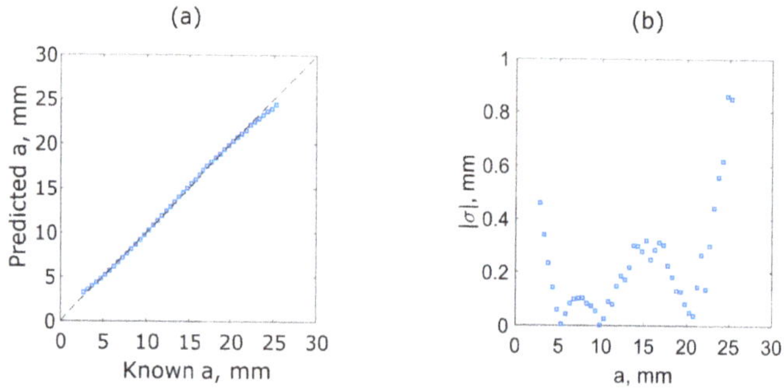

Figure 8. Displacement predictions of bimodal sensors, for co-planar separation, *a*, between solenoid coils. (**a**) The predicted displacement (calculated from experimental measurements and using the fitted functions in Table 2) against the known displacement values. Black dashed lines represent ideal prediction and (**b**) shows the accuracy of the predictions by plotting the absolute error $|\sigma|$ between predicted and known values as a function of linear displacement.

Table 2. Linear regression of the function $y = p_1 x + p_2$ for the displacement of the coils as shown in Figures 6 and 7.

Displacement	y	x	p_1	p_2
a	$e^{\pi k}$	$1/a$	0.39	−0.025
ϕ (sol.)	$e^{8\pi k}$	$1/\phi$	3.98	3.75
ϕ (plan.)	$e^{4\pi k}$	$1/(4-\phi^2)$	81.83	−9.40

The best-fit functions defined by Table 2 were rearranged to produce the best-fit curves (green dashed) for k as a function of the displacement variables as shown in Figures 6c and 7iii. The best-fit curves based on these first-order approximate functions exhibit excellent agreement with the experimental data over a wide range of displacements in all cases. The only exception occurs when $\phi < 10°$ in the solenoid coils, where the assumptions and simplifications of the first-order model break down. Studies were conducted to evaluate the effectiveness of using these functions to predict the physical displacement between sensors based on the experimental measurements of the bimodal resonant frequencies (f_-, f_+).

4.4.1. Evaluating Separation, *a*

Experimental measurements, made every $\Delta a = 1$ mm, were used, and f_+ and f_- were recorded and used to calculate the a using Equation (8) and the best-fit function defined in Table 2. The prediction of a has a limit of $a = 25$ mm, beyond which k becomes negligible and f_+ and f_- are no longer detectable, as demonstrated in Figure 1b,c.

4.4.2. Evaluating Angle, ϕ

The f_+ and f_- peaks were recorded for variations in ϕ in both solenoid and planar coil sensors every $\Delta\phi = 1°$, and the best-fit functions (Table 2) were used to predict ϕ; the results are shown in Figure 9a.

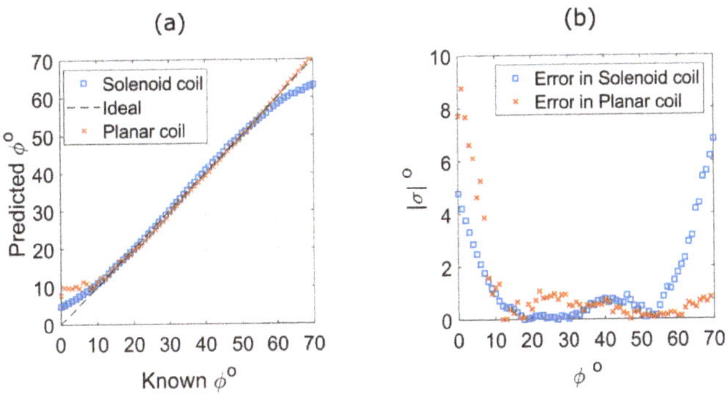

Figure 9. Displacement predictions of bimodal sensors with angular displacement, ϕ, between PCB and solenoid coils. (**a**) shows the predicted displacement (calculated from experimental measurements and using the angular fitted functions in Table 2) against the known displacement values. (**b**) shows the accuracy of the predictions by plotting the absolute error $|\sigma|$ between predicted and known values as a function of angular displacement.

4.4.3. Prediction Error, $|\sigma|$

Figure 8a compares the predicted coil separation, a, using the fitted function in Table 2 compared to that recorded experimentally from the linear translation stage. The absolute error, $|\sigma|$, between the known and predicted values of a is shown in Figure 8b. The results show that the measurement is accurate to within ± 1 mm over the full range of distances measured.

For both solenoid and planar coils, the predictions are more accurate for $\phi > 10°$. At low angles ($\phi < 10°$), the planar sensors do not experience a significant change in k (see Figure 7b(iii)). Hence, the error in the prediction of ϕ is larger for small angles. Conversely, k changes significantly at low angles for the solenoid coils; however, the function employed to fit the solenoid angular displacement has low accuracy at low angles and high angles due to the simplicity of the assumptions employed (see [40]). Additionally, the solenoid coil error increases at angles $\phi > 55°$. This is due to the low gradient of change in the coupling coefficient k preventing accurate interpolation of inverted angles. At these low gradients of change in k, environmental factors are likely to play an important role in the accuracy of angle inversion. These may include thermal fluctuations and background RF interference. However, the most significant source of noise will likely be due to the sensors' proximity to metallic structures, which can cause the resonant frequency of either or both sensors to shift, potentially resulting in a false-positive angular measurement. As shown in Figure 9b, the absolute error in angle for planar sensors decreases to less than $\pm 1°$ for angles $\phi > 10°$. Despite the coil's unoptimised design, the planar coils exhibit excellent prediction capability between 10° and 70°.

5. Conclusions

Pairs of over-coupled inductive coils were used to produce bimodal resonance (frequency-splitting) spectra and resonant frequency tracking was used to evaluate the separation and angular displacement between the coils. The study demonstrated the displacement-dependent bimodal phenomenon in two coil designs experimentally and highlighted the validity of a "centre-of-magnetism" method for defining first-order analytical functions of the coupling interactions between coils; 2D and 3D FE simulations were evaluated, showing excellent agreement with experimental results, and direct inversion of displacement variables showed the technique was able to resolve separations and angles to within ± 1 mm and $\pm 1°$.

While these resolutions are an order of magnitude larger than Hall-effect techniques, the proof of principle in these unoptimised sensor designs demonstrates significant potential for further development of this novel sensing approach. The bimodal resonance tracking method proposed has the potential to be made in real time, deployed directly onto flexible substrates, and expanded to higher degrees of freedom, making it valuable in applications such as robotics, human–computer interaction, and non-destructive testing.

6. Patents

The work presented relates to a pending UK patent application—Inductive Coil Array—UK patent application no. 2215187.2, filed on 14 October 2022.

Author Contributions: Conceptualization, R.R.H.; Methodology, A.H.A., G.O., A.J.M. and R.R.H.; Software, A.H.A.; Validation, A.H.A.; Formal analysis, A.H.A.; Investigation, A.H.A. and G.O.; Writing—original draft, A.H.A.; Writing—review & editing, A.J.M. and R.R.H.; Visualization, A.H.A. and R.R.H.; Supervision, A.J.M. and R.R.H. All authors have read and agreed to the published version of this manuscript.

Funding: Alexis Hernandez's research is funded by the Consejo Nacional de Ciencia y Tecnología (CONACYT) (2020-000008-01EXTF-00426). Robert Hughes's work is funded by EPSRC project Certification for Design—Reshaping the Testing Pyramid (EP/S017038/1).

Institutional Review Board Statement: Not applicable.

Informed Consent Statement: Not applicable.

Data Availability Statement: Data are available at the University of Bristol data repository, data.bris, at https://doi.org/10.5523/bris.nlizgito6ber2vnk8tkyhfl98. Accessed on 9 March 2025.

Conflicts of Interest: The funders had no role in the design of the study; in the collection, analyses, or interpretation of data; in the writing of the manuscript; or in the decision to publish the results. The current employer of the lead author (Ether NDE Ltd.) had no role in the design of the study; in the collection, analyses, or interpretation of data; in the writing of the manuscript; or in the decision to publish the results. Ether NDE Ltd. do provide PhD sponsorship to a different PhD student of author Robert Hughes, on inspection of composite materials.

Appendix A. Derivation of Resonant Frequency of Coupled Coils

Equations (2)–(4) can be used to calculate the impedance spectra, $Z(f)$, for a given coupled system. For the high-frequency case where $R_2 \ll \omega L_2$, the coupling term α from Equation (5) can be simplified to

$$\alpha^2 \approx \frac{L_1}{L_2}\left(\frac{k\gamma_2^2}{\gamma_2^2-1}\right)^2 = \frac{L_1}{L_2}\left(\frac{k\omega^2}{\omega^2-\omega_2^2}\right)^2, \tag{A1}$$

such that the expressions for the effective inductance and resistance in Equations (3) and (4) simplify to

$$L_1' \approx L_1\left[1-\frac{\gamma_2^2}{\gamma_2^2-1}k^2\right], \tag{A2}$$

$$R_1' \approx R_1\left[1+k^2\left(\frac{\gamma_2^2}{\gamma_2^2-1}\right)^2\frac{L_1}{L_2}\frac{R_2}{R_1}\right]. \tag{A3}$$

From Equation (2), by multiplying the numerator and denominator by the complex conjugate of the denominator, we obtain

$$Z_{m1} = \frac{(R_1' + i\omega L_1')\left[(1 - \omega^2 L_1' C_1) - i\omega R_1' C_1\right]}{(1 - \omega^2 L_1' C_1)^2 + \omega^2 R_1'^2 C_1^2}. \quad (A4)$$

The impedance, Z_{m1}, will therefore be a maximum when the denominator is a minimum. Recognising that R_1' only increases with frequency, we can assume that the denominator will be a minimum when

$$1 - \omega^2 L_1' C_1 = 0. \quad (A5)$$

Substituting in Equation (A2) along with the expression for the frequency ratio $r_2^2 = \omega^2 L_2 C_2$ and expanding out the terms, we reach the quadratic equation

$$\omega^4 L_1 C_1 L_2 C_2 \left[1 - k^2\right] - \omega^2 [L_1 C_1 + L_2 C_2] + 1 = 0. \quad (A6)$$

Using the quadratic formula, the general expression for the resonant frequencies of a two-coil system are

$$\omega_0^2 = \frac{1}{2} \frac{\omega_1^2}{[1 - k^2]} \left[1 + \frac{\omega_2^2}{\omega_1^2}\left(1 \pm \frac{\omega_1}{\omega_2}\sqrt{\frac{\omega_2^2}{\omega_1^2} + \frac{\omega_1^2}{\omega_2^2} + 4k^2 - 2}\right)\right], \quad (A7)$$

where $\omega_n = \sqrt{1/L_n C_n}$.

References

1. Yu, X.; Hu, Q.; Xu, D.; Li, P.; Liu, X. Common Sensors in Industrial Robots: A Review. *J. Phys. Conf. Ser.* **2019**, *1267*, 012036. [CrossRef]
2. Melzer, M.; Makarov, D.; Schmidt, O.G. A review on stretchable magnetic field sensorics. *J. Phys. D Appl. Phys.* **2019**, *53*, 083002. [CrossRef]
3. Zhang, J.; Shi, Y.; Huang, Y.; Liang, C.; Dong, Y.; Kang, Y.; Feng, B. A Displacement Sensing Method Based on Permanent Magnet and Magnetic Flux Measurement. *Sensors* **2022**, *22*, 4326. [CrossRef]
4. Jiang, S.; Li, J.; Li, Z.; Li, Z.; Li, W.; Huang, X.; Zhang, H.; Zhang, G.; Huang, A.; Xiao, Z. Experimental realization of exceptional surfaces enhanced displacement sensing with robustness. *Appl. Phys. Lett.* **2023**, *123*, 201106. [CrossRef]
5. Zhang, Z.; Wang, L.; Cao, B.; Zhang, H.; Liu, J. A Moving Magnetic Grid-Type Long-Range Linear Absolute Displacement Sensor. *Sensors* **2023**, *23*, 700. [CrossRef]
6. Zhu, C.; Zhuang, Y.; Liu, B.; Huang, J. Review of Fiber Optic Displacement Sensors. *IEEE Trans. Instrum. Meas.* **2022**, *71*, 7008212. [CrossRef]
7. Berkovic, G.; Shafir, E. Optical methods for distance and displacement measurements. *Adv. Opt. Photon.* **2012**, *4*, 441–471. [CrossRef]
8. Sciammarella, C.A. A Review: Optical Methods That Evaluate Displacement. In *Advancement of Optical Methods & Digital Image Correlation in Experimental Mechanics*; Lamberti, L., Lin, M.T., Furlong, C., Sciammarella, C., Reu, P.L., Sutton, M.A., Eds.; Springer: Cham, Switzerland, 2019; Volume 3, pp. 23–52.
9. Ye, Y.; Zhang, C.; He, C.; Wang, X.; Huang, J.; Deng, J. A Review on Applications of Capacitive Displacement Sensing for Capacitive Proximity Sensor. *IEEE Access* **2020**, *8*, 45325–45342. [CrossRef]
10. Blaž, N.; Kisić, M.; Žlebič, C.; Mišković, G.; Radosavljević, G.; Živanov, L. Displacement sensor based on interdigital capacitor. In Proceedings of the 2015 38th International Spring Seminar on Electronics Technology (ISSE), Eger, Hungary, 6–10 May 2015; pp. 477–481. [CrossRef]
11. George, B.; Tan, Z.; Nihtianov, S. Advances in Capacitive, Eddy Current, and Magnetic Displacement Sensors and Corresponding Interfaces. *IEEE Trans. Ind. Electron.* **2017**, *64*, 9595–9607. [CrossRef]
12. Micro-Epsilon. Precise Distance Sensors. Available online: https://www.micro-epsilon.co.uk/distance-sensors/ (accessed on 2 March 2025).

13. Jagiella, M.; Fericean, S.; Droxler, R.; Dorneich, A. New magneto-inductive sensing principle and its implementation in sensors for industrial applications. In Proceedings of the SENSORS, 2004 IEEE, Vienna, Austria, 24–27 October 2004; Volume 2, pp. 1020–1023. [CrossRef]
14. Jiao, D.; Ni, L.; Zhu, X.; Zhe, J.; Zhao, Z.; Lyu, Y.; Liu, Z. Measuring gaps using planar inductive sensors based on calculating mutual inductance. *Sens. Actuators A Phys.* **2019**, *295*, 59–69. [CrossRef]
15. Ripka, P.; Janosek, M. Advances in Magnetic Field Sensors. *IEEE Sens. J.* **2010**, *10*, 1108–1116. [CrossRef]
16. Tzemanaki, A.; Gao, X.; Pipe, A.G.; Melhuish, C.; Dogramadzi, S. Hand exoskeleton for remote control of minimally invasive surgical anthropomorphic instrumentation. In Proceedings of the 6th Hamlyn Symposium on Medical Robotics, London, UK, 22–25 June 2013.
17. Dupre, N.; Bidaux, Y.; Dubrulle, O.; Close, G.F. A Stray-Field-Immune Magnetic Displacement Sensor with 1% Accuracy. *IEEE Sens. J.* **2020**, *20*, 11405–11411. [CrossRef]
18. Reddy, B.P.; Murali, A.; Shaga, G. Low cost planar coil structure for inductive sensors to measure absolute angular position. In Proceedings of the 2017 2nd International Conference on Frontiers of Sensors Technologies, ICFST 2017, Shenzhen, China, 14–16 April 2017; pp. 14–18. [CrossRef]
19. Owston, C.N. A high frequency eddy-current, non-destructive testing apparatus with automatic probe positioning suitable for scanning applications. *J. Phys. E Sci. Instrum.* **1970**, *3*, 814–818. [CrossRef]
20. Bonfitto, A.; Gabai, R.; Tonoli, A.; Castellanos, L.M.; Amati, N. Resonant inductive displacement sensor for active magnetic bearings. *Sens. Actuators A Phys.* **2019**, *287*, 84–92. [CrossRef]
21. Zhang, Y.; Zhao, Z. Frequency splitting analysis of two-coil resonant wireless power transfer. *IEEE Antennas Wirel. Propag. Lett.* **2014**, *13*, 400–402. [CrossRef]
22. Niu, W.; Jiang, J.; Ye, C.; Gu, W. Frequency splitting suppression in wireless power transfer using hemispherical spiral coils. *AIP Adv.* **2022**, *12*, 055016. [CrossRef]
23. Zhu, P.W.; Wang, X.; Zhao, W.S.; Wang, J.; Wang, D.W.; Hou, F.; Wang, G. Design of H-shaped planar displacement microwave sensors with wide dynamic range. *Sens. Actuators A Phys.* **2022**, *333*, 113311. [CrossRef]
24. Horestani, A.K.; Naqui, J.; Shaterian, Z.; Abbott, D.; Fumeaux, C.; Martín, F. Two-dimensional alignment and displacement sensor based on movable broadside-coupled split ring resonators. *Sens. Actuators A Phys.* **2014**, *210*, 18–24. [CrossRef]
25. Babu, A.; George, B. A linear and high sensitive interfacing scheme for wireless passive LC sensors. *IEEE Sens. J.* **2016**, *16*, 8608–8616. [CrossRef]
26. Babu, A.; George, B. An Efficient Readout Scheme for Simultaneous Measurement from Multiple Wireless Passive *LC* Sensors. *IEEE Trans. Instrum. Meas.* **2018**, *67*, 1161–1168. [CrossRef]
27. Babu, A.; George, B. Sensor System to Aid the Vehicle Alignment for Inductive EV Chargers. *IEEE Trans. Ind. Electron.* **2019**, *66*, 7338–7346. [CrossRef]
28. Ebrahimi, A.; Beziuk, G.; Scott, J.; Ghorbani, K. Microwave Differential Frequency Splitting Sensor Using Magnetic-LC Resonators. *Sensors* **2020**, *20*, 1066. [CrossRef]
29. Schormans, M.; Valente, V.; Demosthenous, A. Frequency Splitting Analysis and Compensation Method for Inductive Wireless Powering of Implantable Biosensors. *Sensors* **2016**, *16*, 1229. [CrossRef]
30. Kalhor, H.A. The Degree of Intelligence of the Law of Biot-Savart. *IEEE Trans. Educ.* **1990**, *33*, 365–366. [CrossRef]
31. Zhang, Y.; Zhao, Z.; Chen, K. Frequency-splitting analysis of four-coil resonant wireless power transfer. *IEEE Trans. Ind. Appl.* **2014**, *50*, 2436–2445. [CrossRef]
32. Hughes, R.; Fan, Y.; Dixon, S. Investigating electrical resonance in eddy-current array probes. *AIP Conf. Proc.* **2016**, *1706*, 090001. [CrossRef]
33. Hughes, R. High-Sensitivity Eddy-Current Testing Technology for Defect Detection in Aerospace Superalloys. Ph.D. Thesis, University of Warwick, Department of Physics, Coventry, UK, 2016.
34. Hughes, R.; Fan, Y.; Dixon, S. Near electrical resonance signal enhancement (NERSE) in eddy-current crack detection. *NDT E Int.* **2014**, *66*, 82–89. [CrossRef]
35. Mazlouman, S.J.; Mahanfar, A.; Kaminska, B. Mid-range wireless energy transfer using inductive resonance for wireless sensors. In Proceedings of the IEEE International Conference on Computer Design: VLSI in Computers and Processors, Lake Tahoe, CA, USA, 4–7 October 2009; pp. 517–522. [CrossRef]
36. Moghaddami, M.; Sundararajan, A.; Sarwat, A.I. Sensorless electric vehicle detection in inductive charging stations using self-tuning controllers. In Proceedings of the 2017 IEEE Transportation Electrification Conference, ITEC-India 2017, Pune, India, 13–15 December 2017; pp. 1–4. [CrossRef]
37. Tyurnev, V.V. Coupling Coefficients of Resonators in Microwave Filter Theory. *Prog. Electromagn. Res. B* **2010**, *21*, 47–67. [CrossRef]
38. Tung, W.S.; Chiang, Y.C.; Cheng, J.C. A new compact LTCC bandpass filter using negative coupling. *IEEE Microw. Wirel. Compon. Lett.* **2005**, *15*, 641–643. [CrossRef]

39. Petrov, P.; Radkovskaya, A.; Shamonina, E. Retrieval of electric and magnetic coupling coefficients. In Proceedings of the 2015 9th International Congress on Advanced Electromagnetic Materials in Microwaves and Optics, METAMATERIALS 2015, Oxford, UK, 7–12 September 2015; pp. 259–261. [CrossRef]
40. Hughes, R.R.; Arroyo, A.H.; Mulholland, A.J. Analytical Approximations for Fitting Magnetic Coupling Coefficients Between Adjacent Coils. *IEEE Trans. Magn.* **2024**, *60*, 8400309. [CrossRef]
41. Wang, Y.; Yao, Y.; Liu, X.; Xu, D.; Cai, L. An LC/S Compensation Topology and Coil Design Technique for Wireless Power Transfer. *IEEE Trans. Power Electron.* **2018**, *33*, 2007–2025. [CrossRef]
42. Zhang, Y.; Lu, T.; Zhao, Z.; He, F.; Chen, K.; Yuan, L. Selective Wireless Power Transfer to Multiple Loads Using Receivers of Different Resonant Frequencies. *IEEE Trans. Power Electron.* **2015**, *30*, 6001–6005. [CrossRef]
43. Yang, D.; Won, S.; Hong, H. Design of Range Adaptive Wireless Power Transfer System Using Non-coaxial Coils. *IOP Conf. Ser. Mater. Sci. Eng.* **2017**, *199*, 012008. [CrossRef]

sensors

MDPI

Article

Magnetic Sensor Array for Electric Arc Reconstruction in Circuit Breakers

Gabriele D'Antona [1,*], Luca Ghezzi [2], Sara Prando [2] and Francesco Rigamonti [2]

[1] Department of Energy, Politecnico di Milano, 20156 Milan, Italy
[2] ABB Electrification, Smart Buildings Division, Vittuone, 20009 Milan, Italy; luca.ghezzi@it.abb.com (L.G.); sara.prando@it.abb.com (S.P.); francesco.rigamonti@it.abb.com (F.R.)
* Correspondence: gabriele.dantona@polimi.it

Abstract: Noninvasive imaging of circuit breakers under short-circuit testing is addressed by recording the magnetic field produced over an array of external sensors and by solving an inverse problem to identify the causing current distribution. The temporal and spatial resolution of the sensing chain are studied and implemented in a physical set-up. A wire model is adopted to describe electrical current distribution. Additionally, the simpler, more direct approach to evaluating the passage of electric current in front of sensors is proposed. The dynamics of suitable approximating models of the electric arc that forms across contacts is obtained and agrees with multi-physical simulations and with experimental time histories of current and voltage. The two methods are flexible and allow the analysis of different types of circuit breakers.

Keywords: circuit breaker testing; short-circuit; current distribution; electric arc; inverse problems; nondestructive testing; magnetic sensors

1. Introduction

Circuit breakers (CBs) are electrical protection devices aimed at sectioning an electric circuit and thus interrupting current flow in the event of a fault that could impair the safety of installations and people. The need to gain a more capillary control in low-voltage installations has led to an increase in the number of circuit breakers (sub-branching) in installations, along with additional types of faults to be detected and cleared. In particular, the traditional protection from overcurrents (i.e., overload and short-circuit), that is found in miniature circuit breakers (MCBs), is now frequently supplemented by residual current protection in the same product, called residual current circuit breakers with overcurrent protection (RCBOs). To keep the size of electrical cabinets within reasonable limits, manufacturers have increased the compactness of circuit breakers. This poses complex challenges to the design of the current interruption means, specifically devoted to extinguishing the hot electric arc plasma that is formed as soon as the electric contacts within the circuit breaker open the circuit.

With reference to Figure 1, the conductive path between terminals <1> and <2> comprises a moving contact <3> that may be driven by a mechanism <4> away from a fixed contact <5>. When a short-circuit fault occurs, the mechanism is actuated by a plunger driven by a solenoid <6>, in series with the current. When an overload occurs, the mechanism is actuated by a bending bimetallic strip <7>, also in series with the current and connected to the moving contact by a flexible braid <8>. The electric arc, originally generated across the contacts, is pushed by Lorentz force towards a rack of splitter plates <9>, so called because they fragment the arc column into a series of many smaller branches. During a successful interruption, a surface voltage across each metal-to-plasma and plasma-to-metal transition builds up a total arc voltage drop comparable with the supply voltage, limiting current and thus causing the arc to quench down. Splitter plates are ferromagnetic to attract the arc. The arc migrates towards the splitters, with its roots running over a pair of metallic rails <10> and <11> termed arc runners and defining a region called the pre-chamber <12>. One arc root has to commute away from the moving contact by jumping onto the arc runner. Back-ignitions are possible if the gaseous atmosphere is still sufficiently hot to propitiate a dielectric breakdown across the contacts.

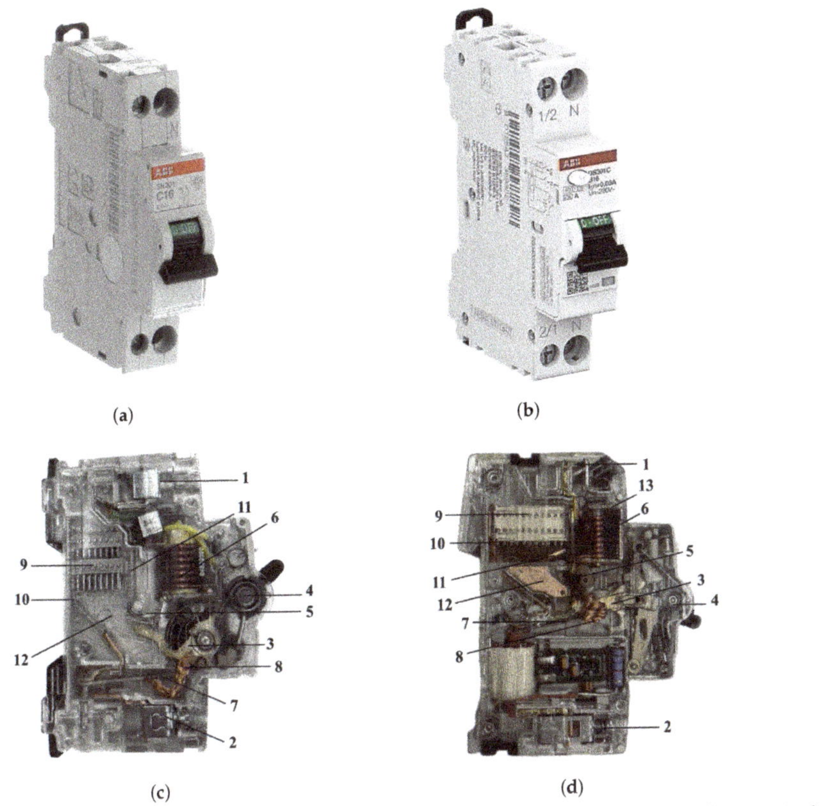

Figure 1. ABB SN201 (**a**) and DS301C (**b**) CBs and their internal structures, (**c**) and (**d**), respectively. See Figure 2 for the size of the arc chamber in the plane of the picture; in the transversal direction, the arc chamber extension is approximately 10 mm.

Arc-based current switching is a complex, nonlinear, multidisciplinary problem. Compressible air flow is governed by the Navier–Stokes equations; Lorentz forces acting upon the conductive plasma column follow the Maxwell equations; the thermal balance that controls arc conductivity heavily depends on radiation with participating media: the gas atmosphere absorbs and scatters heat, plastic and metal solids absorb heat, increase their temperature and may undergo phase transitions, injecting mass and energy in to the gas mixture and releasing droplets that are transported by the flow. Special models address the physics of arc roots; temperature- and pressure-dependent physical properties describe the gaseous atmosphere. The governing equations have been solved numerically by many Authors; see, e.g., Yang et al. [1], Bianchetti et al. [2], Rümpler et al. [3], and references therein. The returned results include flow, temperature, current density and electromagnetic fields anywhere in the computational domain, i.e., inside and outside the circuit breaker under analysis. Understandably, the sophistication of the models calls for experimental validation and model parameter identification, and the stemming complexity amounts to long computational times (days or weeks, depending on the problem size).

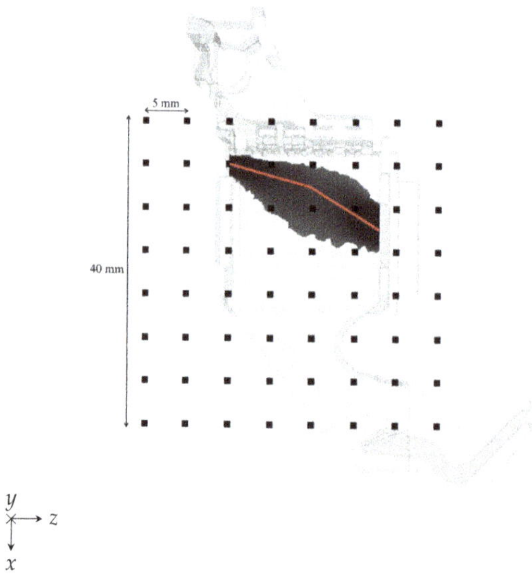

Figure 2. Sample CB measurement configuration and arc current density distribution obtained with the CB MHD model.

Several experimental approaches have been proposed to detect the electric arc dynamics, from arc birth when contacts open to arc extinction, especially in short-circuit conditions. The traditional approach consists of recording the time histories of electrical currents through the circuit breaker and voltage drops across its terminals. Despite the soundness of the approach, the actual shape, size, and spatial distribution of the arc cannot be detected.

Optical methods have been also adopted. Imaging with transparent circuit breaker housings is a common option. Yet, transparent polymers differ from the original polymers that are used in circuit breakers, and the perturbation to the real dynamics poses a strong limit to the effectiveness of the method. Additionally, the trade-off between image resolution and sampling frequency also limits the approach. To overcome the shortcomings, McBride et al. [4] proposed drilling holes in circuit breaker sidewalls and apply optical fibers. However, the preparation process of devices under test is cumbersome. Additionally, the presence of internal parts may obstruct the line of sight of some optical fibers, thus limiting the versatility of the method. Finally, thermal imaging by means of a camera having IR bandpass was evaluated and eventually discarded by Rigamonti [5] and Taccola [6], because the heat flow originating from the hot arc plasma column is overly diffused by plastic sidewalls, resulting in blurred images with poor resolution.

Magnetic arc imaging was proposed by Velleaud et al. [7], with an array of micro coils as field sensors. An advantageous improvement resorting to an array of Hall effect sensors was proposed by Rigamonti [5] and Taccola [6]. In both cases, the magnetic field generated by the arc current is detected by suitable sensors and recorded in time. Then, a mathematical inverse problem is solved on the experimental data set, and the position and simplified shape of the arc column centerline is identified in time.

The present study further develops the approach of Rigamonti [5] and Taccola [6]. Improvements to the experimental setup include to the mechanically supporting structure, which has been redesigned and made more robust and at the same time apt to host multipolar circuit breakers, and a revision of the electronic boards with sensors. More relevant are the improvements to the algorithms. Ferromagnetic inclusions are still modeled by means of a magnetic dipole, as in Rigamonti [5], but the identification of the dipole's property from FEM analysis has been improved. Furthermore, the approach gains in versatility by allowing circuit breakers whose internal current conduction path is subject to change to a second possible configuration along the arc dynamics, owing to discontinuous jumps of arc roots to different portions of the conductive chain. The potentiality of the approach in its improved form has been validated on two circuit breakers, namely the ABB SN201, an MCB, and the ABB DS301C, an RCBO; see Figure 1.

The outline of the rest of this paper is as follows. The requirements and design of the measurement chain are discussed in Section 2, while the resulting experimental setup is briefly discussed in Section 3. In Section 4, the magnetic inversion problem is discussed, with a focus on the improved modeling of ferromagnetic inclusions. Results obtained with the ABB SN201 CB are presented in Section 5, while those relevant to the ABB DS301C CB are in Section 6. In Section 7, the expected accuracy of the method is discussed with reference to results obtained by inverting synthetic data from numerical MHD simulations. Finally, in Section 8, the main conclusions and limitations of the proposed approach are discussed, and opportunities for further research are outlined.

2. Characterization of Detectable Magnetic Signals

Reconstructing arc current density depends heavily on accurately measuring the magnetic field distribution near the circuit breaker. Here, this is achieved by sampling both the temporal and spatial distribution of the magnetic field using a suitable array of magnetic sensors and a data acquisition system. The measurement system must effectively capture and preserve the key signal features. Specifically, it is crucial to specify the following critical signal characteristics:

- Spatiotemporal bandwidth.
- Amplitude range and resolution.

Additionally, proper orientation of the sensor is essential to accurately transduce the magnetic field components.

The specifications of the measuring system were derived considering the reference configuration shown in Figure 2, where the sensors are located on a 40 mm × 40 mm surface at 5 mm from the CB side wall. Magnetic field distributions were evaluated for a 10 kA prospective short-circuit current (RMS value) using a magnetohydrodynamic (MHD) model of the circuit breaker arc. The focus was on the H_y component, as it is significantly influenced by the arc position and shape.

2.1. Time and Space Bandwidth

The time signal bandwidth was evaluated by considering the y-component of the magnetic field at the locations exhibiting the shortest rise time. The energy spectral density (ESD), denoted as $S_{H_y}(f)$, was estimated as the squared magnitude of the signal Fourier transform (see Figure 3).

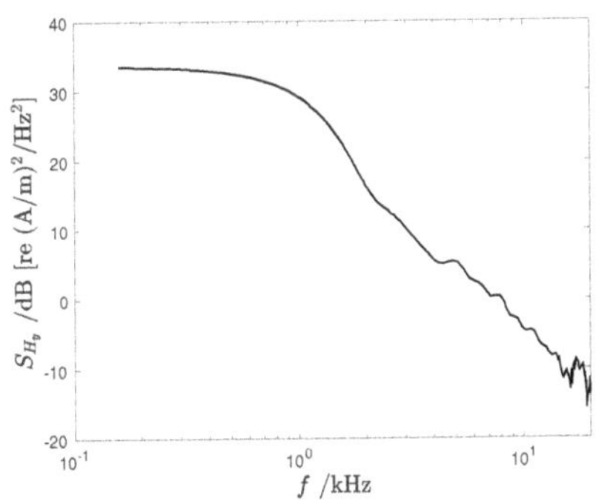

Figure 3. One−sided ESD of the y−component of the magnetic intensity.

The ESD indicates a low-pass signal with a 3 dB bandwidth f_B of approximately 1 kHz and a 40 dB/decade slope in the stop band. Consequently, the ESD was modeled as

$$S_{H_y}(f) = \frac{\kappa}{1 + \left(\frac{f}{f_B}\right)^4}, \quad (1)$$

where $\kappa = 2 \times 10^3\,(\mathrm{A/m})^2/\mathrm{Hz}^2$.

For the measurement chain (comprising the transducer and data acquisition system), it was assumed that the system has a 3 dB bandwidth f_{3dB} with a first-order transfer function given by

$$\underline{G}(f) = \frac{1}{1 + j\left(\frac{f}{f_{3dB}}\right)}. \quad (2)$$

To ensure that the signal energy attenuation is less than 1 %, the 3 dB bandwidth f_{3dB} of the measurement system must satisfy the following inequality

$$\frac{\int_0^{+\infty} |\underline{G}(f)|^2 S_{H_y}(f)\,\mathrm{d}f}{\int_0^{+\infty} S_{H_y}(f)\,\mathrm{d}f} \geqslant 0.99, \quad (3)$$

implying that f_{3dB} must be at least 10 kHz.

To reconstruct the y-component of the magnetic field distribution with limited distortion, it is necessary to define a lower bound to the sampling rate in the spatial x, z domain. This is evaluated using the 2D spatial ESD $S_2(f_x, f_z)$ of the magnetic field distribution. The overall signal energy E_T is given by

$$E_T = \int_{-\infty}^{+\infty}\int_{-\infty}^{+\infty} S_2(f_x, f_z)\,\mathrm{d}f_x\,\mathrm{d}f_z\,. \quad (4)$$

Given f_S as the spatial sampling frequency along the x and z axes, the aliased signal energy E_F is

$$E_F = E_T - \int_{-\frac{f_S}{2}}^{+\frac{f_S}{2}}\int_{-\frac{f_S}{2}}^{+\frac{f_S}{2}} S_2(f_x, f_z)\,\mathrm{d}f_x\,\mathrm{d}f_z\,. \quad (5)$$

Figure 4 shows the ratio of these energies (5) and (4) as a function of the spatial sampling frequency f_S. To ensure E_F is less than 1% of E_T, the energy ratio must exceed 20 dB. Due to the steep slope of the energy ratio around this value, it is advisable to prescribe a spatial sampling frequency of at least 1 cm^{-1}. This condition imposes the upper bound on spatial resolution in the measurement of the magnetic field distribution at

$$d = \frac{1}{1\,\mathrm{cm}^{-1}} = 10\,\mathrm{mm}\,. \quad (6)$$

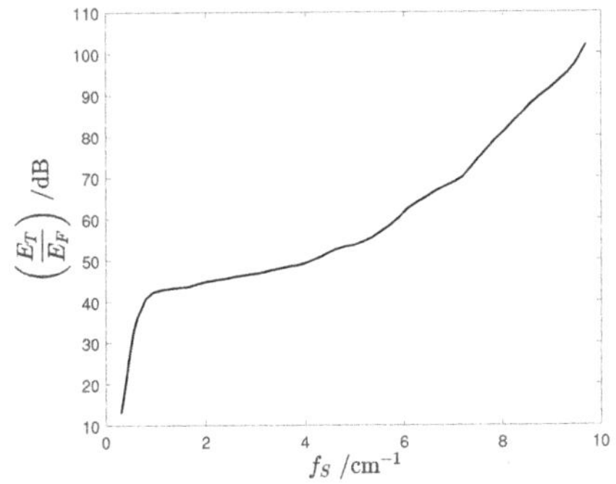

Figure 4. Ratio between the signal energy E_T and the signal aliased energy E_F.

2.2. Signal Dynamic Range

The full scale range for H_y was determined using MHD simulations, which revealed that the magnetic intensity spans approximately to $H_{max} = 25\,\text{kA/m}$. The bound on amplitude resolution of the magnetic measurements is determined by the smallest detectable change in the electric arc current density distribution shape. For simplicity, this bound was deduced using a lumped parameters model instead of the MHD arc model.

The electric arc, as found in the circuit breakers under study, is a plasma cloud without sharp borders. From the analysis of specimens after short-circuit, and particularly from the size of the arc roots, which are clearly evident on metal surfaces touched by the arc feet, the core region of the arc plasma cloud can be estimated to have a diameter of few mm. Multiphysics simulations also confirm this estimate; see, e.g., Figure 20. This inspired the modeling of the arc column as a filament. Despite its coarseness, this wire model approximation fits well with the need to reduce the number of unknowns without jeopardizing excessively the physical shape of the arc. Indeed, the interpretation of the arc filament as the centerline of the plasma cloud succeeds in describing its position relative to the path from electrical contacts to the splitter plates, which matters the most in circuit-breaker development.

The current density is represented as a perturbed straight line filament, resulting from a perturbation of the x-coordinate. The electric arc is described by the parametric line

$$\gamma : \begin{cases} x(u) = \tilde{x} + \delta_m(u) \\ y(u) = \tilde{y} \\ z(u) = L\left(u - \dfrac{1}{2}\right) \end{cases} \qquad \text{with} \quad \delta_m(u) = \begin{cases} \Delta x & m = 0 \\ \sqrt{2}\Delta x \sin\left(m\pi \dfrac{u}{L}\right) & m = 1, 2, \ldots \end{cases} \tag{7}$$

where L is the distance between the arc roots, Δx is the smallest detectable change in the electric arc shape, and $0 \leqslant u \leqslant 1$.

Figure 5 shows the perturbed arc γ for $m = 0, 1, 2,$ and 3. According to the Biot–Savart law, the detectable y-component of the magnetic field variation at $\boldsymbol{r}$ is [8]

$$\Delta H_y(\boldsymbol{r}) = \frac{I}{4\pi}\left(\int_{\gamma} \frac{\mathrm{d}\boldsymbol{l} \times (\boldsymbol{r} - \boldsymbol{x}_\gamma) \cdot \hat{\boldsymbol{y}}}{|\boldsymbol{r} - \boldsymbol{x}_\gamma|^3} - \int_{\gamma_0} \frac{\mathrm{d}\boldsymbol{l} \times (\boldsymbol{r} - \boldsymbol{x}_{\gamma 0}) \cdot \hat{\boldsymbol{y}}}{|\boldsymbol{r} - \boldsymbol{x}_{\gamma 0}|^3} \right), \tag{8}$$

with

$$\mathrm{d}\boldsymbol{l} = \left(\frac{\mathrm{d}x}{\mathrm{d}s}\hat{\boldsymbol{x}} + \frac{\mathrm{d}y}{\mathrm{d}s}\hat{\boldsymbol{y}} + \frac{\mathrm{d}z}{\mathrm{d}s}\hat{\boldsymbol{z}} \right)\mathrm{d}s\,.$$

Here, $\boldsymbol{r}$ is the vector pointing to the sensor position, $\boldsymbol{x}_\gamma$ is the vector spanning the arc line, and $\boldsymbol{x}_{\gamma_0}$ is the vector spanning the unperturbed arc line, denoted γ_0.

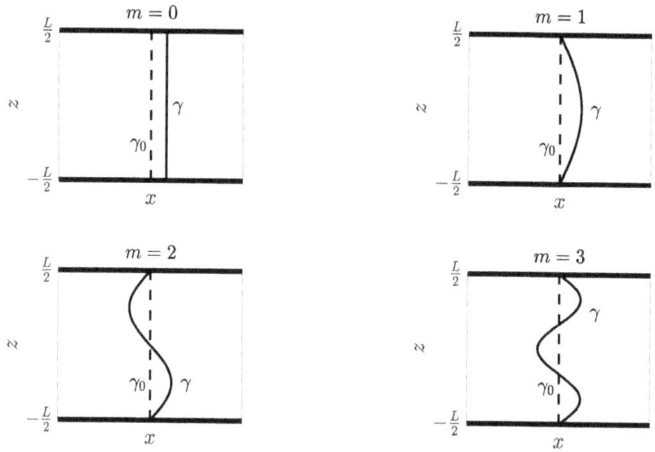

Figure 5. Arc shape perturbations, with $m = 0, 1, 2$, and 3.

Choosing $\Delta x = 1\,\mathrm{mm}$ as the target resolution for addressing the arc shape reconstruction inverse problem, numerical integration of (8) on a plane positioned 5 mm from the CB external sidewall results in peak values of ΔH_y reaching approximately $4\,\mathrm{kA/m}$ for various mode values m. This suggests setting the upper bound on magnetic measurement resolution at

$$\delta H = 1\,\mathrm{kA/m}. \tag{9}$$

2.3. *Sensor Position and Orientation Accuracy*

Position and orientation errors of each sensor affected the measurement accuracy as well. Naming $\boldsymbol{r}$ and $\boldsymbol{r}_N$ the vectors pointing to actual and nominal sensor locations, the position error is (see Figure 6)

$$\Delta \boldsymbol{r} = \boldsymbol{r} - \boldsymbol{r}_N\,. \tag{10}$$

The tilt is defined instead as the difference between the unit vector $\hat{\boldsymbol{n}}$ perpendicular to the sensor tilted surface and the y-axis unit vector $\hat{\boldsymbol{y}}$ (see Figure 6)

$$\Delta \hat{\boldsymbol{n}} = \hat{\boldsymbol{n}} - \hat{\boldsymbol{y}}\,. \tag{11}$$

The magnetic measurement error consequential to positioning and orientation errors is

$$\Delta H = \boldsymbol{H}(\boldsymbol{r}) \cdot \hat{\boldsymbol{n}} - H_y(\boldsymbol{r}_N)\,. \tag{12}$$

Considering the first-order Taylor expansion of the magnetic intensity around the nominal position, we have

$$\boldsymbol{H}(\boldsymbol{r}) \approx \boldsymbol{H}(\boldsymbol{r}_N) + \nabla \boldsymbol{H}(\boldsymbol{r}_N) \cdot \Delta \boldsymbol{r}\,. \tag{13}$$

Using this expansion and Equation (11), the error in the magnetic measurement (12) can be approximated as

$$\Delta H \approx [\boldsymbol{H}(\boldsymbol{r}_N) + \nabla \boldsymbol{H}(\boldsymbol{r}_N) \cdot \Delta \boldsymbol{r}] \cdot (\Delta \hat{\boldsymbol{n}} + \hat{\boldsymbol{y}}) - H_y(\boldsymbol{r}_N)\,. \tag{14}$$

Considering that

$$\nabla \boldsymbol{H}(\boldsymbol{r}_N) \cdot \Delta \boldsymbol{r} \cdot \Delta \hat{\boldsymbol{n}} \approx 0 \quad \text{and} \quad \nabla \boldsymbol{H}(\boldsymbol{r}_N) \cdot \Delta \boldsymbol{r} \cdot \hat{\boldsymbol{y}} = \nabla H_y(\boldsymbol{r}_N) \cdot \Delta \boldsymbol{r}\,,$$

Equation (14) simplifies to

$$\Delta H \approx \underbrace{\boldsymbol{H}(\boldsymbol{r}_N) \cdot \Delta \hat{\boldsymbol{n}}}_{\Delta H_T} + \underbrace{\nabla H_y(\boldsymbol{r}_N) \cdot \Delta \boldsymbol{r}}_{\Delta H_P}\,. \tag{15}$$

Here, ΔH_T represents the error due to the tilt of the sensor, and ΔH_P represents the error due to the positional displacement of the sensor. Thus, the overall magnetic measurement error is the combined effect of these position and tilt contributions. The constraints

$$|\Delta H_P| \leqslant 0.5 \cdot \delta H = 0.5\,\text{kA/m} \quad \text{and} \quad |\Delta H_T| \leqslant 0.5 \cdot \delta H = 0.5\,\text{kA/m} \tag{16}$$

ensure that the total error $|\Delta H|$ remains within the magnetic measurement resolution δH. These constraints determine the bounds on sensor positional displacement $\Delta \boldsymbol{r}$ and tilt $\Delta \hat{\boldsymbol{n}}$.

Using the MHD model, we observe that the largest magnitude (in time and space) of the gradient ∇H_y is

$$\max |\nabla H_y| = 1.6\,\text{A}\,\text{mm}^{-2} \tag{17}$$

and, therefore, the limit for the sensor positional displacement is

$$|\Delta \boldsymbol{r}| < \frac{|\Delta H_P|}{\max |\nabla H_y|} = 0.3\,\text{mm}\,, \tag{18}$$

which can be achieved with both manual and automatic assembly of the sensor array.

At each sensor position, the tilt causes the measurement error

$$\Delta H_T = \boldsymbol{H}(\boldsymbol{r}_N) \cdot (\hat{\boldsymbol{n}} - \hat{\boldsymbol{y}}) = H_x \alpha_x + H_y(\alpha_y - 1) + H_z \alpha_z\,, \tag{19}$$

where

$$\alpha_x = \sin\theta \sin\phi,\ \alpha_y = \cos\theta,\ \alpha_z = \sin\theta\cos\phi \tag{20}$$

are the directional cosines of $\hat{\boldsymbol{n}}$ (Figure 6).

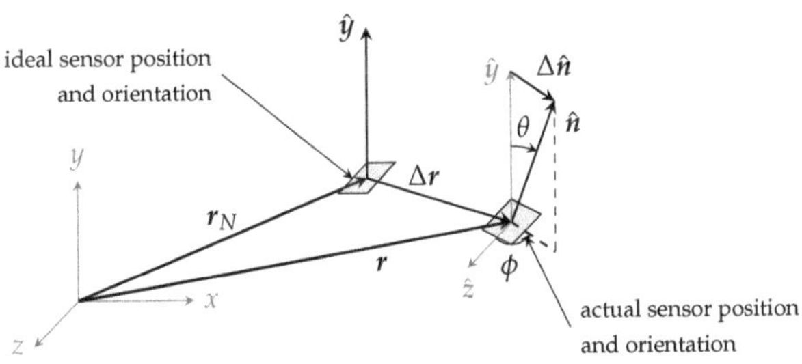

Figure 6. Sensor position error $\Delta \boldsymbol{r}$ and tilt $\Delta \hat{\boldsymbol{n}}$. The y-axis is normal to the ideally untilted sensor surface.

Since (19) depends on numerous interdependent variables, the worst-case error scenario would be overly pessimistic and unlikely. Therefore, to assess the uncertainty stemming from tilt, a statistical approach is employed. This involves treating all relevant quantities as random variables, and checking if the root mean square value of ΔH_T complies with bound (16). The mean square value $m^2_{\Delta H_T}$ of ΔH_T, linked to the tilt induced uncertainty in each magnetic field measurement, reads

$$m^2_{\Delta H_T} \approx j \begin{bmatrix} \boldsymbol{R}_H & \boldsymbol{0} \\ \boldsymbol{0} & \boldsymbol{R}_\alpha \end{bmatrix} j^T\,, \tag{21}$$

where

- j is the Jacobian vector of (19) with respect to the components of $\boldsymbol{H}$ and the directional cosines of $\hat{\boldsymbol{n}}$, evaluated at their mean values (hereinafter represented by a bar over the symbols) and given by

$$j = \begin{bmatrix} \overline{\alpha}_x & (\overline{\alpha}_y - 1) & \overline{\alpha}_z & \overline{H}_x & \overline{H}_y & \overline{H}_z \end{bmatrix}; \tag{22}$$

- $\boldsymbol{R_H}$ is the auto-correlation matrix of the magnetic intensity components;
- $\boldsymbol{R_\alpha}$ is the auto-correlation matrix of the directional cosines of $\hat{\boldsymbol{n}}$.

The mean values of the directional cosines are derived under the assumption that θ and ϕ in Equation (20) are mutually independent and uniformly distributed with zero mean. Therefore,

$$\overline{\alpha}_x \approx \sin\overline{\theta}\sin\overline{\phi} = 0, \quad \overline{\alpha}_y \approx \cos\overline{\theta} = 1, \quad \overline{\alpha}_z \approx \sin\overline{\theta}\cos\overline{\phi} = 0\,. \tag{23}$$

Thus, the first three elements of the Jacobian vector (22) are zero. Consequently, (21) reduces to

$$m^2_{\Delta H_T} \approx \begin{bmatrix} \overline{H}_x & \overline{H}_y & \overline{H}_z \end{bmatrix} \boldsymbol{R_\alpha} \begin{bmatrix} \overline{H}_x \\ \overline{H}_y \\ \overline{H}_z \end{bmatrix}. \tag{24}$$

The mean values of the magnetic field components, estimated by processing the data from the MHD model, are as follows:

$$\overline{H}_x = -0.5\,\mathrm{kA/m}, \quad \overline{H}_y = -0.4\,\mathrm{kA/m}, \quad \overline{H}_z = -7.6\,\mathrm{kA/m}\,. \tag{25}$$

The auto-correlation matrix $\boldsymbol{R_\alpha}$ of the directional cosines in (20) is derived assuming that the angle ϕ ranges from $-\pi$ to π, while the maximum value for the angle θ is computed as (refer to Figure 7)

$$\theta_{max} = \tan^{-1}\left(\frac{t_{max}}{w}\right).$$

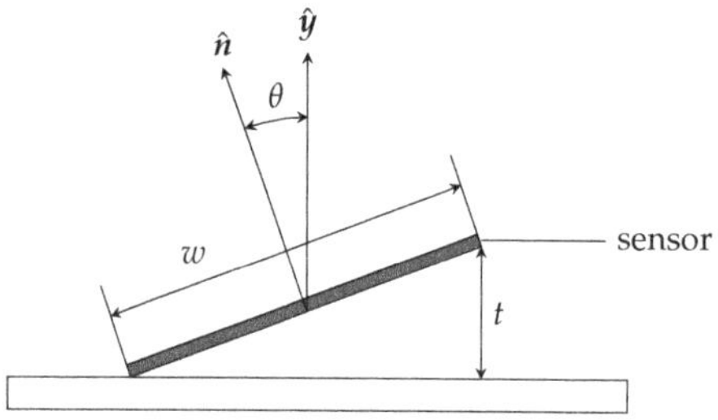

Figure 7. Sensor tilt angle θ.

In the case of manually assembled sensors, and assuming that the sensor are manufactured as surface-mount devices with minimum size $w = 2\,\mathrm{mm}$, the maximum misalignment t_{max} is of the order of 0.2 mm, resulting in a tilt angle upper bound

$$\theta_{max} = \tan^{-1}\left(\frac{t_{max}}{w}\right) = 0.10\,\mathrm{rad}\,. \tag{26}$$

The auto-correlation matrix $\boldsymbol{R_\alpha}$, estimated with 10^6 Monte Carlo trials for manual assembly, is

$$\boldsymbol{R_\alpha} = \begin{bmatrix} 1.7 & 0.0 & 0.0 \\ 0.0 & 1000 & 0.0 \\ 0.0 & 0.0 & 1.7 \end{bmatrix} 10^{-3}\,. \tag{27}$$

In the case of automatically assembled sensors, the maximum misalignment can be reduced to 50 µm, leading to the tilt angle upper bound

$$\theta_{max} = \tan^{-1}\left(\frac{t_{max}}{w}\right) = 0.03\,\text{rad}\,. \tag{28}$$

For automatic assembly, the auto-correlation matrix reads

$$\boldsymbol{R}_{\alpha} = \begin{bmatrix} 0.15 & 0.00 & 0.00 \\ 0.00 & 1000 & 0.00 \\ 0.00 & 0.00 & 0.15 \end{bmatrix} 10^{-3}\,. \tag{29}$$

According to (21), the root mean square errors are:

- In the case of manual assembly, $m_{\Delta H_T} = 0.52\,\text{kA/m}$.
- In the case of automatic assembly, $m_{\Delta H_T} = 0.43\,\text{kA/m}$.

Given the constraint $|\Delta H_T| \leqslant 0.5\,\text{kA/m}$ as per inequality (16), only automatic assembly meets the requirements.

3. Sensor Array Layout and Data Acquisition System

Table 1 summarizes the magnetic measurement requirements derived in the previous section. Considering the measuring range B_{max} specified in Table 1 and bearing market availability in mind, along with size and ease of acquiring the output signal, Hall effect sensors are the preferred option. For commercial devices with an integrated electronic conditioning circuit, the noise level declared by the manufacturers is of the order of 100 µT over a 15 kHz bandwidth, which complies with requirements. Still compliant, though higher and more worrying than the noise level, is the linearity error, which can be of the order of 0.5% to 1% of the sensor range (i.e., 0.5 mT to 1 mT, considering ±100 mT for the sensor span) [9,10].

Table 1. Measurement system requirements.

Parameter	Symbol	Value
FS Range	$B_{\text{max}} = \mu_0 H_{max}$	⩾31 mT
System bandwidth	f_{3dB}	⩾10 kHz
Spatial resolution	d	⩽10 mm
Amplitude resolution	$\delta B = \mu_0 \delta H$	⩽1.3 mT

In particular, we focused our choice on the Honeywell Hall effect sensor SS39ET (subminiature SOT-23 surface mount package) [11]. The dimensions of this Hall effect sensor are 1.90 mm × 1.60 mm × 1.20 mm ($L \times W \times T$). These dimensions make this sensor suitable for assembling a matrix meeting the spatial resolution requirement d in Table 1. The other characteristics that make this sensor suitable for this application are

- Full scale: 100 mT > B_{max};
- Linearity error: 0.7 mT < δB;
- 3 dB bandwidth: 115 kHz > $f_{3\text{dB}}$;
- Output voltage span: 0.95 V to 4.05 V, (5 V supply voltage);
- Sensitivity: 14 mV/mT.

The Hall effect sensors were arranged in an array consisting of eight rows and eight columns, forming a square matrix totaling 64 sensors, covering a 40 mm × 40 mm area. This layout ensures a magnetic field distribution measurement with a resolution of 5 mm, compliant with the specified limit resolution d. To achieve a reliable and easy-to-assemble system, each of the eight rows was implemented as a PCB (denoted "strip") with eight sensors (refer to Figure 8a). The complete array was then reconstructed by aligning these strips side by side. A plastic slotted frame was designed to facilitate this assembly process, depicted in Figure 8b.

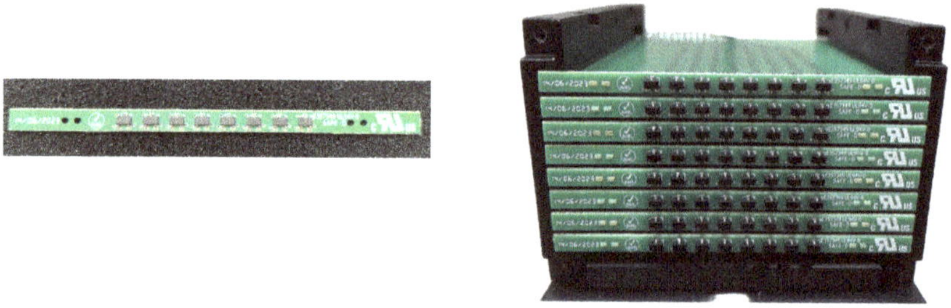

(**a**) Sensor strip board detail. (**b**) Sensors array.

Figure 8. Layout of the sensor array.

The sensor array is positioned outside the CB at a specific distance to protect the sensors from hot gases or high temperatures generated during short-circuits. This positioning is illustrated in Figure 2 and depicted in Figure 9.

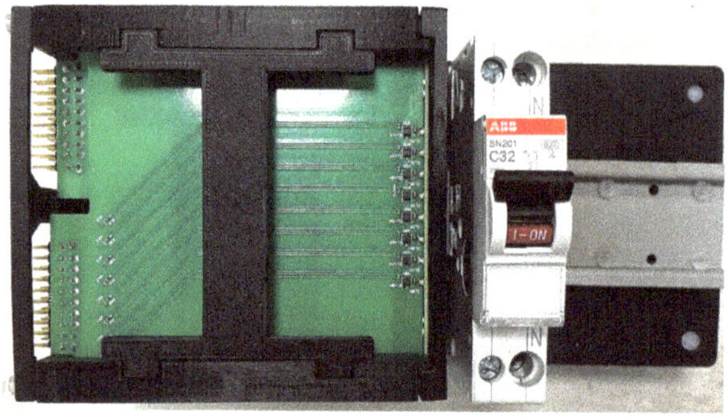

Figure 9. Sensors array with a CB under test.

The connection of the 64 sensors to the transmission lines, which link the sensor array to the data acquisition system, could not be established at the sensor position due to the excessive magnetic induced voltages picked up during the short-circuit transient.
To mitigate this issue, a two-layer PCB was designed to be placed orthogonally to the sensor plane, as the magnetic field intensity decreases in the direction orthogonal to the circuit breaker surface. This design reduces magnetic coupling with the cabling connectors, and transmission line wires; hence, the term *vertical* boards is used hereinafter.

Considering the Hall effect sensor sensitivity, along with the required measurement resolution and bandwidth, experimental and simulated current transient data were used to determine that a vertical PCB with a length between 50 mm and 150 mm would meet the required voltage pickup requirement. Additionally, it was established that the distance between the two signal traces for each sensor of the array should be kept less than 1 mm. The dimensions of the vertical boards, shown in Figure 10a, are 80 mm × 90 mm.

Given the limited ability of the Hall effect sensors to reject power supply disturbances picked up by the sensor power supply line on the vertical PCB, a fast voltage regulator was placed as close to each sensor as possible, as a farther resource supplementing the use of a stabilized voltage source for the sensor array (see Figure 10b). Additionally, on the opposite side of the vertical PCB, near the transmission line connectors, an RC passive low-pass filter was installed on each sensor channel (see Figure 10a). These filters have a cutoff frequency of 48 kHz, effectively reducing the system bandwidth and the noise, while still being compliant with the specified bandwidth bound $f_{3\,\mathrm{dB}}$.

To accommodate a total of 65 channels, comprising 64 for magnetic signals and 1 dedicated to short-circuit current, a PXI bus system was developed using the National Instruments PXIe-8133 chassis. This system integrates five National Instruments PXIe-6358 boards, each capable of simultaneous sampling. Each PXIe-6358 board offers 16 differential analog inputs with 16-bit resolution, a sampling rate of 1.25 MHz, and a full-scale input range of −5 V to 5 V. To facilitate the connection to these boards, shielded I/O connector blocks (SCB-68 model) were employed. These blocks simplify the handling of connectors and also facilitate the installation of required bias resistances for each differential input channel. The sensitivity $S = 14\,\mathrm{mV/mT}$ of the Hall effect sensors and the magnetic field resolution $\delta B = 1.3\,\mathrm{mT}$ dictate a voltage measurement resolution given by

$$\delta V = S \cdot \delta B = 14\,\mathrm{mV/mT} \cdot 1.3\,\mathrm{mT} = 18.2\,\mathrm{mV}\,. \tag{30}$$

The resolution of the PXIe-6358 boards, which have a 16-bit resolution (N) and a full-scale (FS) input of 5 V, meets this requirement. The resolution is calculated as

$$\frac{2 \cdot FS}{2^N - 1} = \frac{2 \cdot 5\,\mathrm{V}}{2^{16} - 1} = 0.2\,\mathrm{mV} < \delta V\,. \tag{31}$$

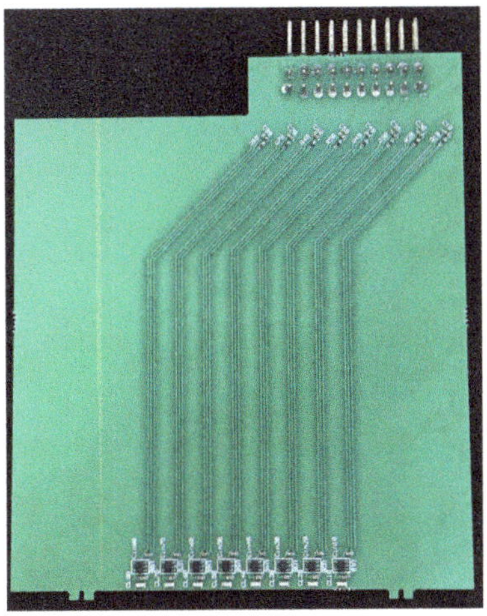

(**a**) Vertical PCBs.

(**b**) Sensor connections on vertical PCBs.

Figure 10. Vertical PCBs and sensor connections.

4. Inverse Magnetic Problem

In the following, a unique Cartesian reference frame (O, x, y, z) is adopted to describe space, and the inverse problem is implicitly referred to a generic time instant t of interest. Magnetic field measurements are available in an array of M sensor locations $\boldsymbol{r}_k \in \mathbb{R}^3$, for $k \in \{1, \ldots, M\}$. The magnetic field measured by the sensors is due to two sources: electric currents and induced magnetization currents in ferromagnetic inclusions (splitter plates).

A *wire model* is used for electric currents. The conductive path is modeled as a network $(\mathcal{E}, \mathcal{V})$ with E connected, conductive edges and V vertices; see Figures 12 and 16. Edges represent filaments, or wires, that connect vertices; they are here represented as straight segments, for simplicity, though generalizations to other sorts of curves could be straightforwardly considered. The modeling network's edge set $\mathcal{E}$ is partitioned into a subset $\mathcal{E}_m$ corresponding to the branch inclusive of the moving contact (blue in figures), a subset $\mathcal{E}_r$ corresponding to the two arc runners (green and pink), a subset $\mathcal{E}_a$ corresponding to the electric arc (red), and a subset $\mathcal{E}_f$ corresponding to all of the remaining edges, which are fixed during the current interruption process (black). A current I_i is associated with the ith edge. In the cases analyzed, the edge network contains no more than two branches in parallel and the total current I is known from measurements. Let I_a be the current flowing through the parallel branch inclusive of the electric arc. Then, let $\beta := I_a/I$ be the (unknown and to be identified) corresponding fraction over the total current I. Then, depending on its (known) location in the network, the ith edge e_i carries a current

$$I_i = \begin{cases} \beta I & \text{if } e_i \in \mathcal{E}_m, \\ (1-\beta)I & \text{if } e_i \in \mathcal{E}_a \bigcup \mathcal{E}_r, \\ I & \text{if } e_i \in \mathcal{E}_f. \end{cases} \tag{32}$$

For $j \in \{1, \dots, V\}$, let $\boldsymbol{v}_j \in \mathbb{R}^3$ be a position vector storing the coordinates of the jth vertex. An E-by-2 connectivity matrix $\mathbf{C} = [c_{ij}]$ describes how edges connect, i.e., c_{i1} (resp., c_{i2}) is the number of the first (resp., second) vertex of the ith edge e_i. Let $u \in [0,1]$ span the generic ith edge. Correspondingly, $\boldsymbol{x}_i(u) := (1-u)\boldsymbol{v}_{c_{i1}} + u\boldsymbol{v}_{c_{i2}}$ is a point on that edge ($\boldsymbol{x}_i(0) = \boldsymbol{v}_{c_{i1}}$ and $\boldsymbol{x}_i(1) = \boldsymbol{v}_{c_{i2}}$). The vector from tail to head of edge e_i is $\boldsymbol{v}_{c_{i2}} - \boldsymbol{v}_{c_{i1}}$, while $(\boldsymbol{v}_{c_{i2}} - \boldsymbol{v}_{c_{i1}})/\|\boldsymbol{v}_{c_{i2}} - \boldsymbol{v}_{c_{i1}}\|_2$ is the corresponding unit vector. Multiplying the latter by du, one gets an infinitesimal vector along the direction of edge e_i, and further multiplication by I_i yields an infinitesimal current vector along edge e_i. This vector represents the infinitesimal current at $\boldsymbol{x}_i(u)$, while $\boldsymbol{r}_k - \boldsymbol{x}_i(u)$ represents the distance vector from the latter location to the kth sensor $\boldsymbol{r}_k$. Then, the magnetic field contribution at sensor $\boldsymbol{r}_k$ due to electric currents is computed integrating over all edges with the Biot–Savart formula, which is exact (in the approximation of currents as wires) and reads [8]

$$\boldsymbol{H}_c(\boldsymbol{r}_k) = \frac{1}{4\pi} \sum_{i=1}^{E} \int_0^1 I_i \frac{(\boldsymbol{v}_{c_{i2}} - \boldsymbol{v}_{c_{i1}}) \times (\boldsymbol{r}_k - \boldsymbol{x}_i(u))}{\|\boldsymbol{v}_{c_{i2}} - \boldsymbol{v}_{c_{i1}}\|_2 \|\boldsymbol{r}_k - \boldsymbol{x}_i(u)\|_2^3} du. \tag{33}$$

The integrand in (33) is a quotient of polynomials and radicals of polynomials. As such, it is integrated exactly, with elementary techniques.

Magnetization currents are modeled by means of a magnetic dipole $\boldsymbol{m} = [m_x, m_y, m_z] \in \mathbb{R}^3$, located in a position $\boldsymbol{x}_m = [x_m, y_m, z_m] \in \mathbb{R}^3$ and representative of the whole rack of splitter plates. The choice of a single magnetic dipole representative of the whole stack of splitter plates is justified by the smoothness of magnetic field $\boldsymbol{H}$ at the sensor locations, as indicated by magnetic FEM simulations. In fact, given the distance from the splitters to the sensors, the magnetic effects on sensor boards are not dissimilar from an equivalent, unique, bulky ferromagnetic inclusion embedding the splitter plates. The magnetic effects of finer structures such as splitter plate thickness and relative spacing are damped out at the sensor locations, which also justifies the greater spacing of sensors compared to that of the splitter plates. Recalling that $\boldsymbol{r}_k - \boldsymbol{x}_m$ is a distance vector from the location of the magnetic dipole to the kth sensor $\boldsymbol{r}_k$, the magnetic field contribution in the latter location due to magnetization currents (in the magnetic dipole approximation) reads [8]

$$\boldsymbol{H}_m(\boldsymbol{r}_k) = \frac{1}{4\pi} \left(3(\boldsymbol{r}_k - \boldsymbol{x}_m) \frac{\boldsymbol{m} \cdot (\boldsymbol{r}_k - \boldsymbol{x}_m)}{\|\boldsymbol{r}_k - \boldsymbol{x}_m\|_2^5} - \frac{\boldsymbol{m}}{\|\boldsymbol{r}_k - \boldsymbol{x}_m\|_2^3} \right). \tag{34}$$

The total magnetic field at sensor $\boldsymbol{r}_k$ is then $\boldsymbol{H}(\boldsymbol{r}_k) = \boldsymbol{H}_c(\boldsymbol{r}_k) + \boldsymbol{H}_m(\boldsymbol{r}_k)$. The sensor read-out is the orthogonal component $h_k^{\perp} := \boldsymbol{H}(\boldsymbol{r}_k) \cdot \hat{\boldsymbol{n}}$, where $\hat{\boldsymbol{n}} \in \mathbb{R}^3$ is the normal unit vector to the plane of sensors. The M magnetic field values at sensor locations can be collected in a vector $\boldsymbol{h} = [h_k^{\perp}] \in \mathbb{R}^M$.

The parameters to be identified, namely the quantities that uniquely define the electric current network and the magnetization dipole (for brevity, the sources), can be collected in a vector $\boldsymbol{p} \in \mathbb{R}^N$. Correspondingly, a suitable operator $\mathcal{L} : \mathbb{R}^N \to \mathbb{R}^M$ is built by computing (33) and (34) and the other expressions defined above, such that

$$\boldsymbol{h} = \mathcal{L}(\boldsymbol{p}). \tag{35}$$

The *direct problem* consists of computing the magnetic field values $\boldsymbol{h}$ from the sources $\boldsymbol{p}$, and admits a unique solution. The *inverse problem* consists of estimating the sources $\boldsymbol{p}$ from the measures of magnetic field values $\boldsymbol{h}$, and is solved by minimizing a suitable discrepancy $\Gamma(\boldsymbol{p})$ between the reference measures $\boldsymbol{h}$ and the corresponding model predictions $\mathcal{L}(\boldsymbol{p})$. Following Rigamonti [5], the discrepancy reads

$$\Gamma(\boldsymbol{p}) := \|\boldsymbol{h} - \mathcal{L}(\boldsymbol{p})\|_2^2 + \alpha\ell(\boldsymbol{p}), \tag{36}$$

where the 2-norm $\|\boldsymbol{h} - \mathcal{L}(\boldsymbol{p})\|_2$ accounts for the distance between measures and model predictions, while, as a regularization term, $\ell : \mathbb{R}^N \to \mathbb{R}_+$ is an operator that computes the length of the arc column.

Inverse problems are usually ill-conditioned, meaning that little perturbations in input data (here, magnetic field measures, for instance, due to experimental noise or other errors) may result in non-small perturbations in output data (here, arc reconstruction). Intuitively, the direct problem is naturally regularized by the smoothing effect of source-to-sensor distance; see Biot–Savart law (33) and magnetic dipole approximation (34). Therefore, the inverse problem tends to be ill-conditioned. Hence, the need for the second term on the r.h.s. of (36), implementing the classical Tikhonov regularization technique. A penalty factor $\alpha \geqslant 0$ weighs the relevance of the regularization term. Increasingly higher values of Tikhonov parameter α penalize arc bends, pushing towards a rectilinear shape. As in [5], the actual value of α has been found numerically, with the so-called L curve technique, i.e., by repeatedly inverting synthetic data from a MHD simulation in correspondence with different values of α and choosing the minimizer of the reconstruction error; see, e.g., [12] for a theoretical reference. In the cases under study, a plateau in a comfortably large neighborhood of the selected value $\alpha = 0.2$ leads to an easy choice. A new calibration would realistically be needed if the inverse problem structure should change significantly, for instance, if more sensors and/or unknowns are introduced, or if different kind of circuit breakers are addressed.

The sources $\boldsymbol{p}$ are constrained to belong to an admissible domain Ω, reflecting the physical bounds on their values. As mentioned, the identified sources are

$$\tilde{\boldsymbol{p}} := \arg\min_{\boldsymbol{p}\in\Omega} \Gamma(\boldsymbol{p}). \tag{37}$$

The solution is found numerically by means of the Levenberg–Marquardt algorithm [13].

The actual expression of the $\mathcal{L}$ operator depends on the choice of parameters. Two different stages may be defined. First, for each different type of circuit breaker, the magnetic dipole is identified independently from the current reconstruction problem. Assuming the breaker contacts are in the closed position and a current intensity in the short-circuit range, a magnetostatic, FEM analysis is run to compute the magnetic field at the sensor locations, accounting for the presence of ferromagnetic inclusions that are magnetized by induction; see Figure 11. The electrical contacts are closed and there is no arc, so the penalty on its length may be conveniently neglected by setting $\alpha = 0$. Since the current path is completely defined and known, the only parameters to be identified are

$$\boldsymbol{p}_{df} = [x_m, y_m, z_m, m_x, m_y, m_z]. \tag{38}$$

Let $\mathcal{L}_{df}$ be the form taken by operator $\mathcal{L}$ for $\boldsymbol{p} = \boldsymbol{p}_{df}$, that is, the full description of the magnetic dipole, and let $\tilde{\boldsymbol{p}}_{df}$ be the corresponding solution to (37). The inverse problem has 64 Equations (8 by 8 sensor array) and 6 unknowns. Physical intuition suggests that the magnetic dipole should approximately lie parallel to the splitter plates and be approximately located midway in direction z along the stack thereof. Consequently, Rigamonti [5] assumed a priori the value of z_m and set $m_z = 0$. Rigamonti focused mainly on symmetrically shaped splitter plates, in which case the dipole is located in the middle of the plate in direction y, thus fixing a priori also the value of y_m (and setting $m_x = 0$). In this study, in addition to the full characterization (38), a reduced characterization

$$\boldsymbol{p}_{dr} = [x_m, m_x, m_y] \tag{39}$$

is also considered, taking inspiration from the simplified treatment in Rigamonti [5]. The inverse problem has 64 equations and 3 unknowns. While $\boldsymbol{p}_{dr}$ was identified in [5] on a trial and error basis by comparison with magnetostatic FEM analyses, in this study $\mathcal{L}_{dr}$ is the form taken by operator $\mathcal{L}$ for $\boldsymbol{p} = \boldsymbol{p}_{dr}$, that is, the reduced description of the magnetic dipole, and $\tilde{\boldsymbol{p}}_{dr}$ is the corresponding solution to (37). A good agreement between Rigamonti's assumptions and the full dipole characterization is found, also for non symmetric splitter plates, like in the case of the ABB DS301C CB; see Figure 11, where the asymmetric splitter plate is outlined in black. The second identification problem is the actual current reconstruction. The magnetic dipole is now characterized either in full or in reduced form. In either case, its location $\boldsymbol{x}_m$ and its direction $\hat{\boldsymbol{u}}_m := \boldsymbol{m}/\|\boldsymbol{m}\|_2$ are known, while its magnitude m, such that $\boldsymbol{m} = m\hat{\boldsymbol{u}}_m$, is a free parameter to be identified. This way, the actual dependence of magnetization currents dynamics from the electric current intensity is retained. Physical reasons constrain m. In particular, $|m| \leqslant M_s V$, where M_s is a material-dependent, maximal magnetization corresponding to magnetic saturation, and V is the volume of ferromagnetic inclusions. As for electric currents, two curvilinear abscissae U_1 and U_2 describe the position of the arc roots on the two arc runners, and dynamically redefine the extension of $\mathcal{E}_r$ to be considered in (33). The arc $\mathcal{E}_a$ is dynamically described by a 2-edge poly-line having as vertices the two roots and a point having the same distance from the latter and defined by means of the (signed) displacement U_3 from the midpoint of the segment connecting the arc roots. The parameters U_1, U_2, U_3 have the dimension of length. An angle ψ, to be identified, locates the moving contact, and thus defines $\mathcal{E}_m$ and characterizes a stretchable edge connecting a predetermined point on the moving contact to a predetermined point on the fixed contact. The parameter vector, therefore, reads

$$\boldsymbol{p}_c = [U_1, U_2, U_3, \psi, \beta, m], \tag{40}$$

$\mathcal{L}_c$ is the corresponding form taken by operator $\mathcal{L}$, and $\tilde{\boldsymbol{p}}_c$ is the corresponding solution to (37). The inverse problem has 64 equations and 6 unknowns. Parameter β (identified) and total current I (directly measured), along with (32), allow reconstructing all edge currents, at each time instant.

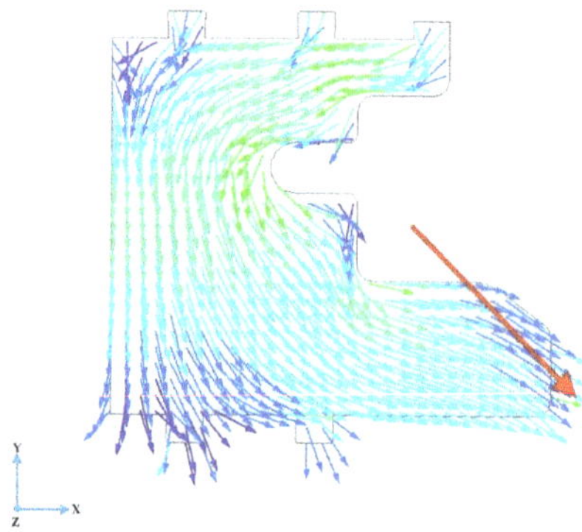

(**a**) (**b**)

Figure 11. ABB DS301C: (**a**) Actual splitter plates stack; (**b**) Model of splitter plate (black outline), magnetic dipole $\boldsymbol{m}$ (red, tail point is $\boldsymbol{x}_m$), and magnetic flux density, computed numerically with magnetostatics FEM (in false colors). Figures 15 and 19 show side views of the rack of splitter plates, where the reconstructed arc-approximating poly-line is also visible.

5. Analysis of ABB SN201 MCB

The first analyzed device is an ABB SN201 MCB (Figure 1). The conductive path of the device is shown in Figure 12 along with its 3D wire model. The short-circuit tests performed on the circuit breaker under analysis were realized in the phase-to-ground condition in order to generate one electric arc only in the phase's extinction chamber. To maximize arcing time and consider a severe fault, the short-circuit insertion angle and the RMS prospective current value were set to 45° and 4.5 kA, respectively. The experiments took place in the ABB short-circuit laboratory in Vittuone, Italy. The sensor array plane is 2 mm away from the device's sidewall to ensure that the sensors do not saturate and can detect the magnetic flux properly. Suitable plastic shims were used to guarantee that this distance was maintained during the test. The circuit breaker was firmly mounted on the plastic DIN rail, ensuring that the system remains fixed despite the strong electrodynamic forces.

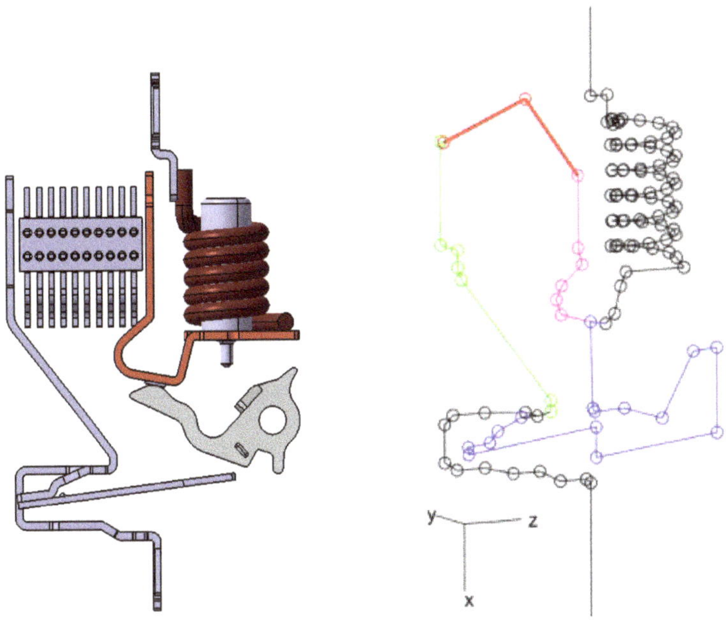

Figure 12. ABB SN201 case: geometry of the conductive chain and wire model circuit. The blue edges represent the branch including the moving contact, while the green and pink edges depict the paths of the two arc runners. The red edges signify the electric arc, and the black edges denote all remaining fixed components during the current interruption process.

Figure 13a shows the magnetic signals for one of the central sensor strips of the array: since these Hall effect sensors are located in the central area of the extinction chamber <12>, the corresponding signals vary their signs depending on the position of the electric arc. The oscillogram in Figure 13b reports the voltage and current time evolution during the phenomenon taken into account for the resolution of the inverse magnetic problem.

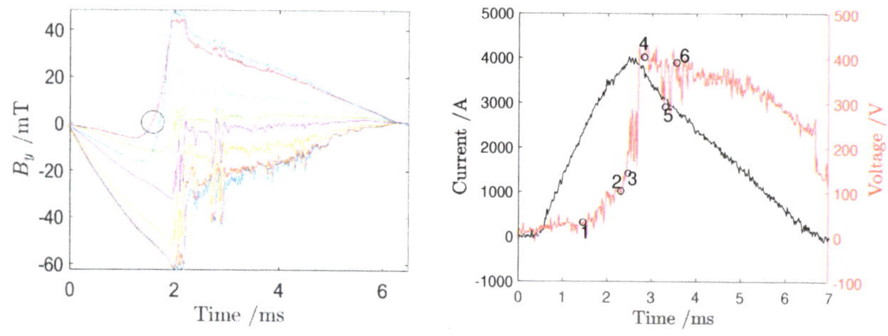

(**a**) (**b**)

Figure 13. (**a**) Magnetic signals from the central strip, with each sensor's signal shown in a different color. The first zero−crossing instant is circled in black. (**b**) Current and voltage data acquired using the short−circuit laboratory's acquisition system.

5.1. Zero-Crossing Method

As previously highlighted, the magnetic signals vary their sign depending on the position of the sensor with respect to the electric arc: a signal experiencing zero-crossing means that the arc is passing over the corresponding sensor. By evaluating these zero-crossing instants, an initial, rough reconstruction of the electric arc's evolution is obtained. This allows a first comprehension of what happens inside the arc chamber during the short-circuit extinction. The phenomenon can be summed up as follows, referring to Figure 14: At first, the magnetic signals follow the short-circuit current magnitude trend, then the mobile contact opens and the electric arc starts forming. The arc then moves inside the extinction chamber towards the splitter plate area. The noisy part of the signals indicates the electric arc's movement inside the chamber, the arc's fragmentation and possible restrikes if the voltage and gas temperature levels induce dielectric breakdown across the contacts. Lastly, the magnetic signals decay, indicating arc extinction. By overlapping the sensor grid to the internal structure of the device under analysis, the map of the zero-crossing sensors at different time ranges is easily obtained. Figure 14 illustrates the estimated evolution of the arc exploiting the zero-crossing method.

The presented method allows the identification of the main initial movements of the electric arc but it has a limited spatial resolution. Hence, the need for solving the aforementioned inverse magnetic problem and reconstructing properly the arc's position at any given instant.

5.2. Inverse Magnetic Problem

Figure 13b highlights salient instants during the interruption process (1 to 6). The short-circuit starts at $t = 0$; the instant the voltage exhibits a first step up, at around 2 ms, indicates that the electric arc is being ignited between the fixed contact <5> and the moving contact <3>, which is opening. The voltage builds up quickly meaning that the arc is now formed and is migrating towards the pre-chamber <12>. From 3 ms onwards, the voltage is consistently high, indicating the splitter plates' partial usage as the arc moves towards them. From this instant up to the final arc extinction, the electric arc plasma will mainly be distributed inside the splitters <9>.

Figure 15 shows a visual representation of the evolution of the electric arc in the aforementioned instants. It shows the reconstructed poly-line overlapped with the actual structure of the device. The reconstruction of the salient moments of arc evolution agrees with the zero-crossing method: The magnetic signals follow the short-circuit current magnitude trend, instant (1) (Figure 15a); the mobile contact opens and electric arc starts forming, instant (2) (Figure 15b); the arc then moves inside the extinction chamber towards the splitter plate area, instants (3) (Figure 15c) and (4) (Figure 15d). In addition, arc restrike is now detected and is shown in Figure 15e, instant (e). Then, the electric arc decays, instant (6) (Figure 15f). The figure also presents the current magnitude in false colors.

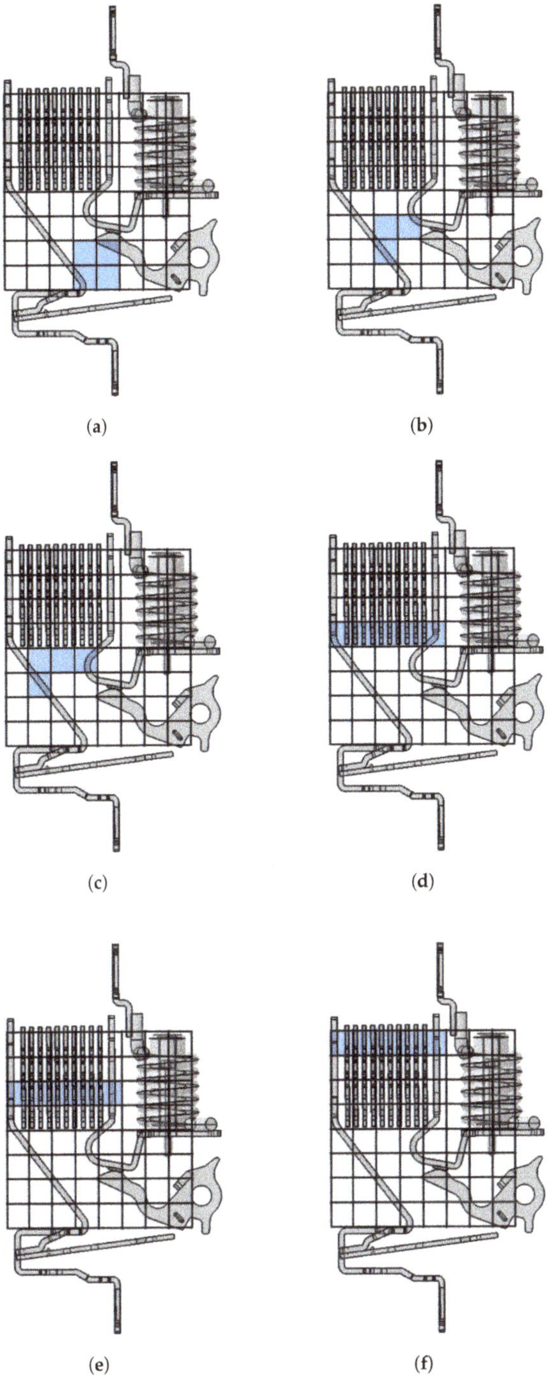

Figure 14. Results of 4.5 kA short-circuit test with ABB SN201 using the zero-crossing method [14]. The letters (**a**–**f**) indicate successive moments in time, showing the progression during the initial phase of the short–circuit.

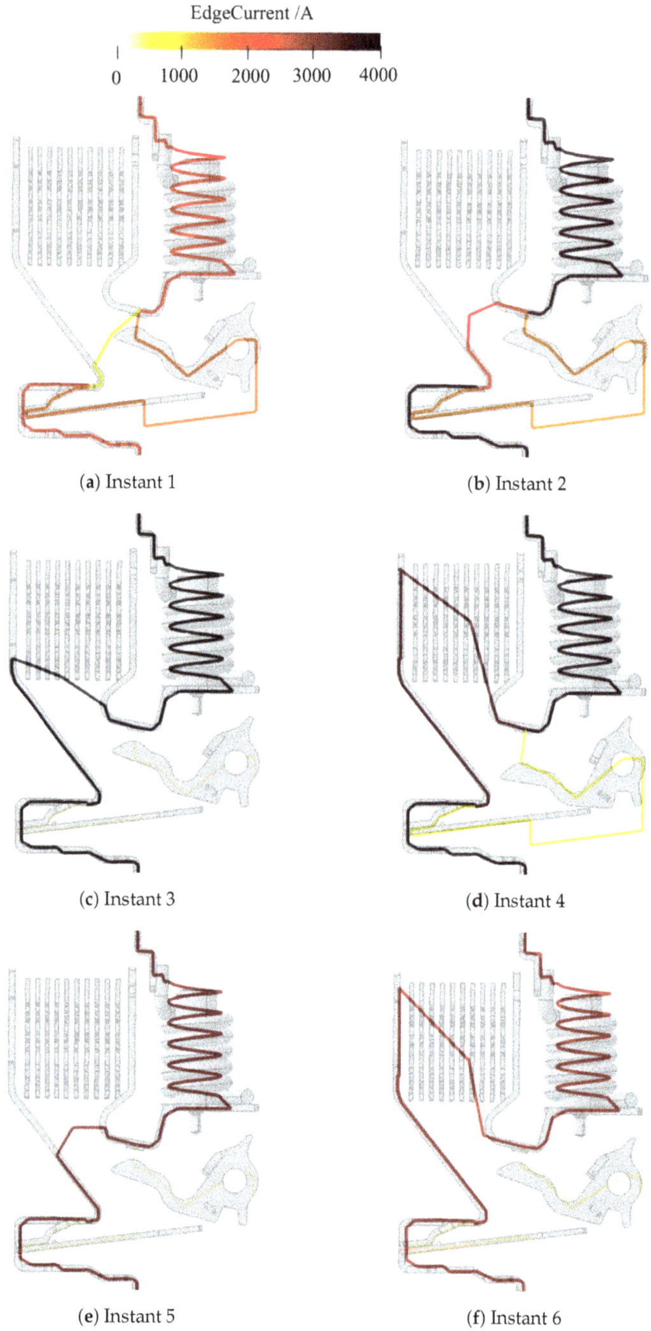

Figure 15. Results of experimental data inversion of ABB SN201 [14]. Same time indexing as Figure 13b.

6. Analysis of ABB DS301C RCBO

The structure of the ABB DS301C RCBO differs from the previously considered circuit breaker: it presents lateral arc plates in the pre-chamber <12> and has a magnetic yoke <13> enclosing the short-circuit coil <6>, both of which are visible in Figure 1d. Arc plates on both sides of the circuit breaker are introduced to direct the electric arc towards the splitter plate area <9> more quickly, in particular at low currents, around 1–2 kA. Their presence would have shielded the phenomenon during the acquisition process: the issue is solved by removing them. The presented study concerns, thus, a modified version of the ABB DS301C.

The wire model of the considered RCBO differs from that of the previously studied CB. The arc root along the rail connected to the fixed contact <5> may reside on the hook <11> or the yoke <13>. The arc root may reside in either of the two portions, which determines one of the two 3D wire models shown in Figure 16. The parameter identification process is run over the two configurations, and the solution taken is associated with the lowest discrepancy (36). The magnetic yoke <13> is intended as a magnetic flux concentrator for the short-circuit trip coil actuator. Indeed, it locally increases the magnetic force on a movable ferromagnet inside the trip coil solenoid. Yet, the magnetic effects outside of the short-circuit trip coil, and particularly in the arc chamber and at the sensor locations, are minor and thus, they are neglected in this study. In case other ferromagnetic structures should play a non-negligible role, one could use the same approach as for splitter plates, i.e., an additional magnetic dipole approximation. Clearly, this would increase the number of parameters to be identified.

In Figure 17b, instants until (5) are resolved with the arc root on the hook, while (6) corresponds to a jump to the yoke, which happens in the final phase of the extinction process. The results agree with expectations. Further evidence comes from the actual arc root erosion as per the visual inspection performed after the short-circuit test. The other current path in the yoke around the solenoid <6> actually lacks continuity and it is, thus, not considered. The same test conditions as in the case of the ABB SN201 CB are chosen for the ABB DS301C CB except for the distance of the RCBO sidewall from the sensors array, which is 3 mm in this case.

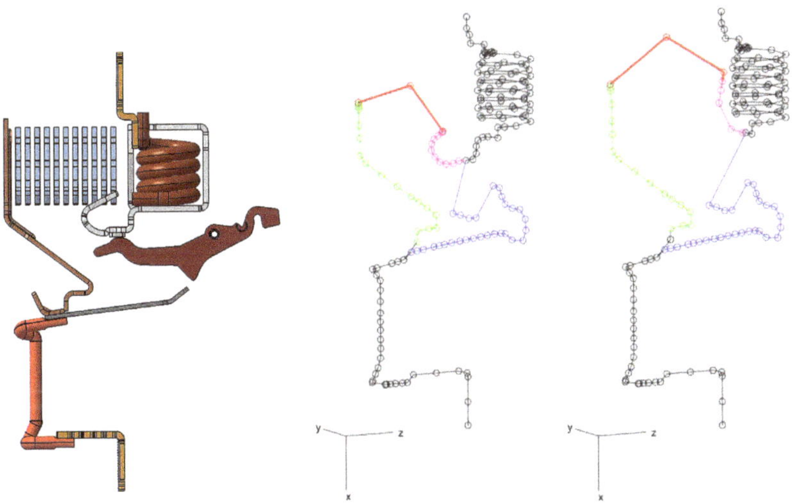

Figure 16. ABB DS301C case. From left to right: geometry of the conductive chain, wire model circuit in case of arc root on the hook and the lower part of the yoke. The blue edges represent the branch including the moving contact, while the green and pink edges depict the paths of the two arc runners. The red edges signify the electric arc, and the black edges denote all remaining fixed components during the current interruption process.

Figure 17a shows the magnetic signals for one of the central sensor strips of the array: since these Hall effect sensors are located in the central area of the extinction chamber <12>, the corresponding signals vary their signs depending on the position of the electric arc. The oscillogram in Figure 17b shows the voltage and current time evolution during the phenomenon taken into account for the resolution of the inverse magnetic problem.

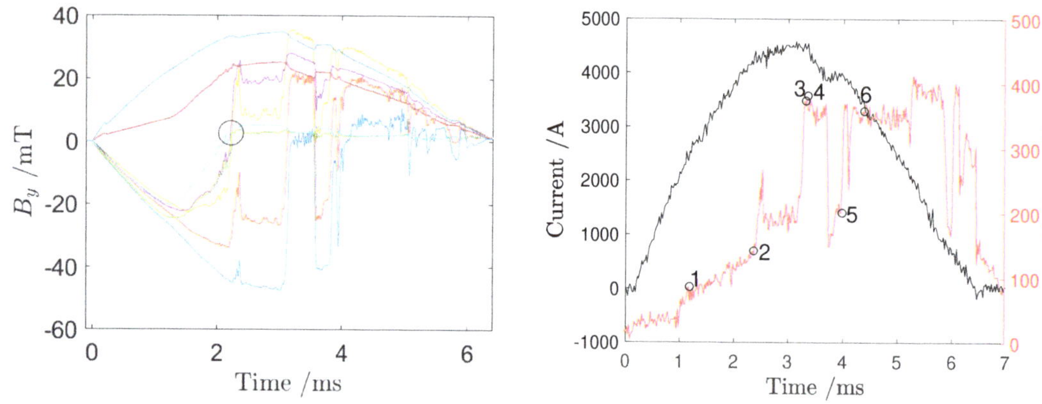

(a) (b)

Figure 17. (**a**) Magnetic signals from the central strip, with each sensor's signal shown in a different color. The first zero−crossing instant is circled in black. (**b**) Current and voltage data acquired using the short−circuit laboratory's acquisition system.

6.1. *Zero-Crossing Method*

Similarly to the ABB SN201 case, a first reconstruction of the evolution of the phenomenon is possible by tracking the instants when the sensor signals cross zero. The results agree with expectations; see Figure 18.

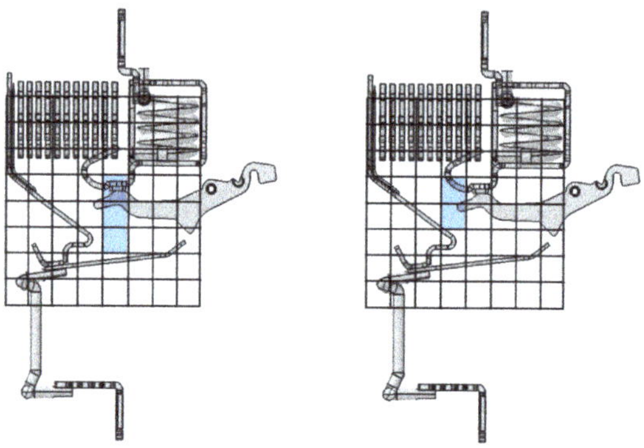

(a) (b)

Figure 18. *Cont.*

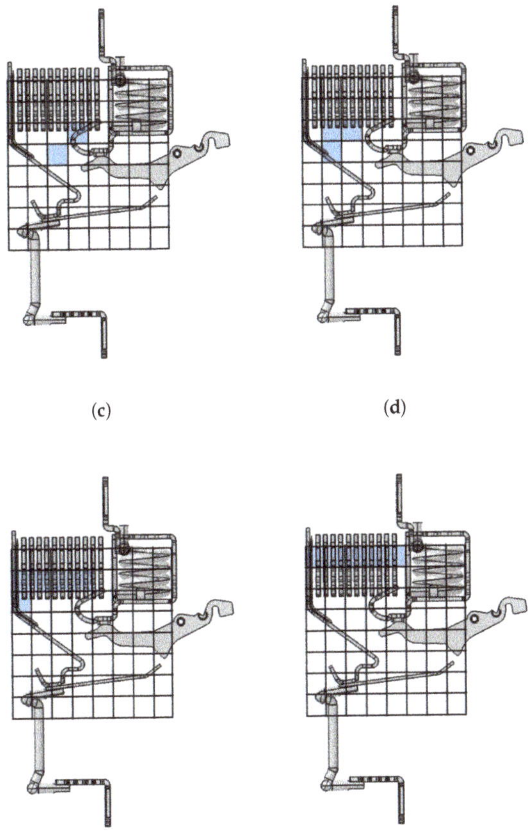

(c) (d)

(e) (f)

Figure 18. Results of 4.5 kA short-circuit test with ABB DS301C using the zero-crossing method [14]. The letters (**a–f**) indicate successive moments in time, showing the progression during the initial phase of the short–circuit.

6.2. Inverse Magnetic Problem

The oscillogram in Figure 17b highlights the values of the voltage corresponding to some instants taken into account for the final inversion (1 to 6). Figure 19 reports the visual representation of the electric arc evolution in the aforementioned instants. It shows the poly-line overlapped with the actual structure of the device. The reconstruction of the salient moments of arc evolution agrees with the zero-crossing method. In addition, arc restrike in the pre-chamber <12> is now detected and is reported in Figure 19e. The last frame (Figure 19f) shows the passage of the arc root from the hook to the yoke enclosing the coil. The electric arc is now fully inside the splitter plates area where it will quench down until the final arc extinction.

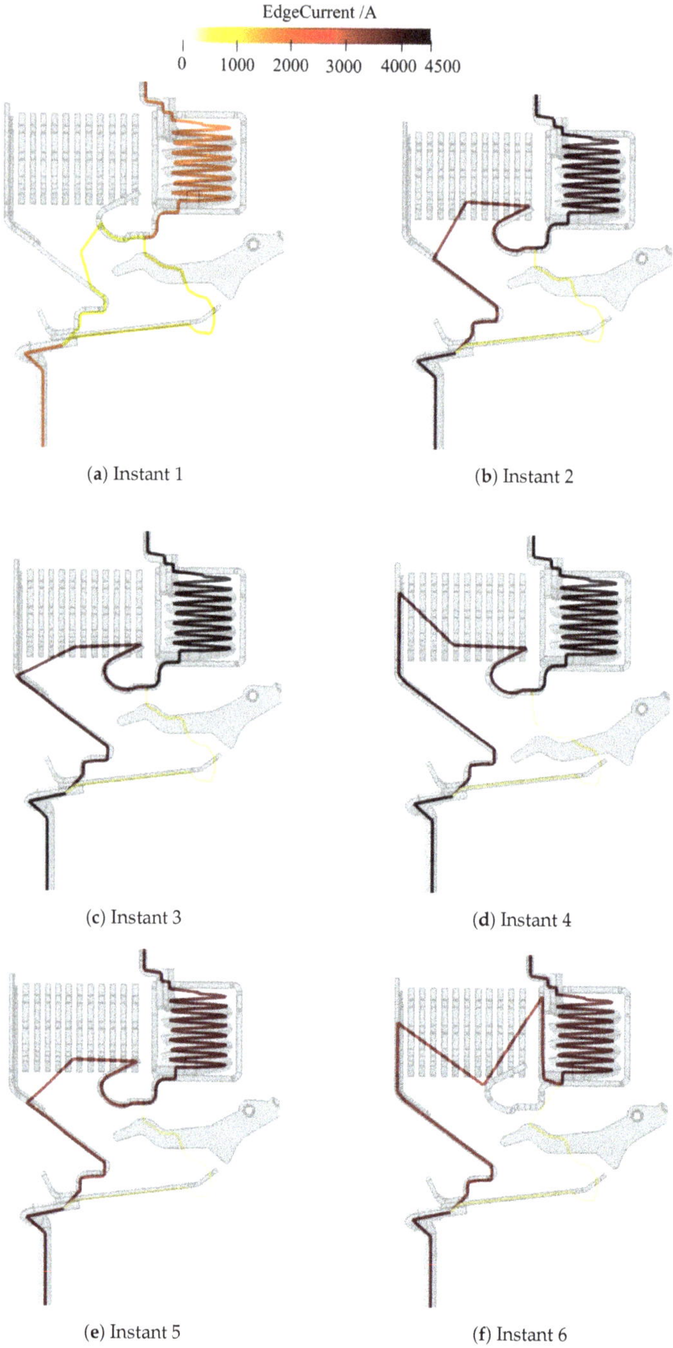

Figure 19. Results of experimental data inversion of ABB DS301C [14]. Same time indexing as Figure 17b.

7. Solution Quality Assessment

It is important to discuss the impact of data errors on the solution. A comprehensive analysis of the impact of various error sources on the reconstructed arc current is very complex and beyond the scope of this paper. There are two main unresolved issues that contribute to this complexity.

Firstly, defining the error in the inverse problem solution is challenging. The problem of comparing the 3D arc current density distribution with the poly-line that represents the wire current model obtained from the inverse problem is complex and does not have a unique solution. In [5], by drawing an analogy between mass density distribution and current density distribution, the solution error was defined as the average distance between the poly-line current solution and the principal axis of inertia of the current density.

However, a drawback of this simple error definition is that it concentrates the current distribution along a straight segment (the principal axis of inertia), losing information about the bending of the current distribution in the plasma. A better approach would be to define the error as the average distance between the poly-line current solution and the line representing the locus of current density centroids in planes orthogonal to the principal axis of inertia.

In [5], another simpler approach was considered, referred to as the Center of Current Density (CoCD) method, which takes advantage of the arc chamber geometry. The region between the rails is cut with three parallel planes, and the current density centroid is computed for each plane. The five points obtained (three from the planes and two from the rails) are then used to approximate the current density distribution with a poly-line. This method is simpler than the previously proposed method. However, in the case of different breaker geometries, the location of the cutting planes could be ambiguous or impractical.

Additionally, Ref. [5] demonstrated that regularization mitigates the effect of data errors on the solution error. It was also shown that even when both data errors and regularization are removed, the error in the inverse problem solution does not reduce to zero. This residual error represents the effect of approximating a current density distribution as a wire current. For the MCB SN201, Ref. [5] found that this error contribution is on the order of 1 mm.

Secondly, while data errors are typically modeled as random vectors, with their effects on the solution statistically described using second-order moments (covariance and correlation matrices), the regularization applied to the inverse problem is deterministic. Integrating the probabilistic representation of data errors with the deterministic constraints of prior regularization remains an unresolved issue in inverse problem research. Although a Bayesian reinterpretation has been proposed to address this [15,16], it is not directly applicable in this context. Moreover, expressing the model error introduced by representing the arc as a filament current instead of a 3D current density distribution in probabilistic terms adds further complexity. This further complicates the ability to express the expected error affecting the estimated arc poly-line shape in actual laboratory experiments, where the true arc current density distribution is unknown. This topic remains an open question that warrants further research and dedicated publications.

To assess the quality of the inverse problem solution presented in this paper, a complete inversion is performed on synthetic data generated by the MHD arc simulation tool developed at ABB. Using this approach, the arc freely evolves in the arc chamber, with its shape, size, path, and movement governed by MHD equations. While full simulation involves substantial computational effort, the tool is used as a synthetic data generator. This simulation tool provides a highly accurate representation of magnetic fields outside the circuit breaker during a short-circuit, accounting for a realistic arc shape resulting from its natural, free evolution. Sample results relevant to the ABB SN201 MCB are displayed in Figure 20, for two different stages of the arc migration from contacts to splitter plates. The arc clouds from the MHD synthetic data are superimposed to the poly-lines (with three points in between arc roots) obtained by solving the corresponding inverse problems. Visual inspection shows a satisfactory agreement, qualitatively assessing the capability of the proposed method in detecting the position of the arc centerline in the frame of the arc chamber, and particularly with reference to the proximity to or entrance into the splitter plate rack.

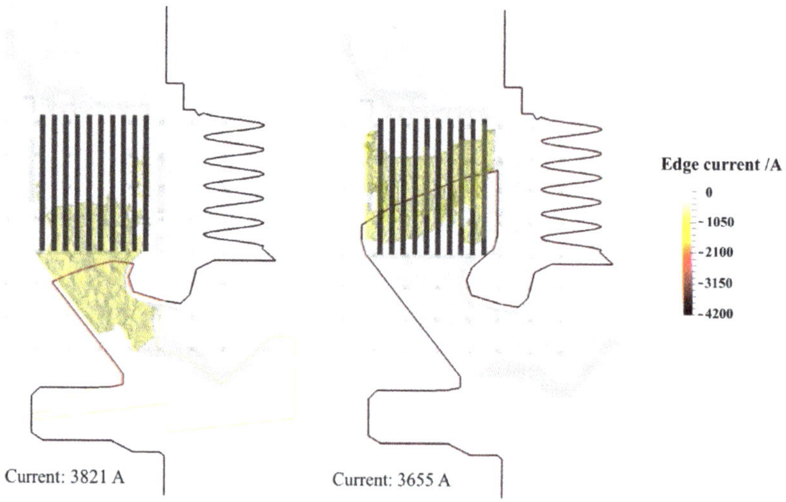

Figure 20. Reconstruction of synthetic data computed numerically with ABB MHD arc simulation tool on an ABB SN201 MCB; The wire model current obtained by the inverse problem approach is superimposed on the arc cloud−shaped current density distribution.

8. Conclusions

Two diagnostic methods have been considered: zero-crossing and resolution of the inverse magnetic problem. Despite its simplicity, the zero-crossing method returns a valuable reconstruction of the arc dynamics. This does not imply the resolution of an inverse problem and, therefore, it is not affected by constraints on the number of mathematical unknowns that describe the arc shape, for its fineness is mainly due to the number of sensors. This allowed dealing with the complex geometry of the ABB DS301C circuit breaker in a straightforward manner. Both methods agree with one another and with the oscillograms, available simulations, physical intuition, and inspection of samples after the short-circuit tests.

The two proposed arc diagnostic methods return a believable time history of the arc centerline position. The relevance of this piece of information resides in the crucial fact that a well-performing circuit breaker is characterized by arcs that quickly and steadily migrate from the electrical contacts to the splitter plates, where they are extinguished. In contrast, poorly performing circuit breakers are characterized by arcs that may hesitate before entering the splitters, or that may restrike back in between electrical contacts or in the pre-chamber. It is, therefore, apparent that the proposed approach catches the salient information for beaker developers. The experimental set-up and the algorithms are thought of for supplementing ordinary short-circuit oscillograms (i.e., current and voltage time histories). Since the approach is experimental, the prospected usage is a posteriori: design variations are implemented in prototypes and tested, and the magnetic diagnostics improve the physical insight from oscillograms. The hardware is found to be robust enough to coexist with the severe physical conditions of a short-circuit chamber, where high currents produce high magnetic fields, Lorentz forces, and strong over-voltages, and where soot, dust, and fumes from burnt polymers are common. The algorithms are simple enough to run quickly, without slowing down the testing operations.

The proposed method, at the present stage of research, exhibits some limitations. Short-circuits involving phase and neutral, or multiple phases, have yet to be addressed.

The number of unknowns necessary to describe multiple arcs would realistically require more measurement points. Furthermore, the distance between magnetic field sources (i.e., currents) and sensors worsens the conditioning of the inverse problem, because changes in the sources are damped out. Thus, arcs burning in-between contacts that are farther from the sensor array plane are more challenging to identify. Nonetheless, in the frame of circuit-breaker engineering, the crucial test is the single-pole short-circuit to ground. Therefore, the practical impact of the above shortcomings is minor. The geometrical characterization of the arc column is admittedly rather crude, but it may be improvable if more sensors are introduced, to gain higher realism. The miniaturization trend in electronic components could possibly yield smaller sensors, that could be mounted closer one another, and perhaps with 3-axis magnetic field read-out. Revising the hardware in the short to mid term is, therefore, a possibility. If bigger circuit breakers than MCBs and RCBOs are to be examined, the whole set-up should be reconsidered, at least for the practical reason of adapting it to a different geometry. Similarly, the location and orientation of the sensor array is a consequence of the quasi-planar geometry of the conductive path, and particularly of the electric arc. More elaborated geometries would require a study to define the optimal location and orientation of sensors so that they still measure the most informative component of magnetic field $\boldsymbol{H}$. If, practical difficulties could arise, such as the physical construction of sensor boards, nonetheless such modifications comply with the nature of the proposed approach, so that the theory and algorithms should remain applicable. Yet, in more general contexts than those addressed so far, there could exist state variables, describing the electrical arc system, that are not observable, i.e., it could be impossible to estimate them from the available measures. A dedicated study would be necessary to check the feasibility of the proposed approach. The bigger size of breakers suited for higher currents offers the opportunity to install many more sensors and enrich the dimension of the measurement vector, which in turn could lead to more refined arc modeling. However, the cost of the data acquisition system would increase accordingly.

This study starts from the work of Rigamonti [5] and Taccola [6]. The virtue of the proposed method is validated and reinforced by the analysis of further circuit breakers. Future improvements to the method may pertain to the number of sensors and the algorithm. In the case of the former, a second array of similar sensors, located on the opposite sidewall of the CB, could provide a higher resolution and could allow the recording of two simultaneous arcs, as in the case of a short-circuit from phase to neutral (as opposed to phase to ground). In the case of the inverse problem algorithm, different description of the arc shape may be attempted, or, in correspondence of arc root commutations, the choice between different wire models may be organized on different grounds.

Author Contributions: Conceptualization, G.D. and L.G.; methodology, G.D., L.G., S.P. and F.R.; software, S.P. and F.R.; validation, S.P. and F.R.; formal analysis, S.P. and F.R.; investigation, S.P. and F.R.; resources, L.G.; data curation, S.P. and F.R.; writing—original draft preparation, G.D., L.G. and S.P.; writing—review and editing, G.D., L.G. and S.P.; visualization, S.P. and F.R.; supervision, G.D. and L.G.; project administration, G.D. and L.G.; funding acquisition, L.G. All authors have read and agreed to the published version of the manuscript.

Funding: This research received no external funding.

Institutional Review Board Statement: Not applicable.

Informed Consent Statement: Not applicable.

Data Availability Statement: Data is contained within the article. The original contributions presented in the study are included in the article, further inquiries can be directed to the corresponding author/s.

Conflicts of Interest: Author Luca Ghezzi, Sara Prando, and Francesco Rigamonti were employed by the company ABB S.p.A. The remaining authors declare that the research was conducted in the absence of any commercial or financial relationships that could be construed as a potential conflict of interest.

References

1. Yang, F.; Wu, Y.; Rong, M.; Sun, H.; Murphy, A.; Ren, Z.; Niu, C. Low-voltage circuit breaker arcs—Simulation and measurements. *J. Phys. D Appl. Phys.* **2013**, *46*, 273001. [CrossRef]
2. Bianchetti, R.; Adami, A.; Fagiano, L.; Gati, R.; Hofstetter, L. Arc Simulation in Low Voltage Switching Devices, a Case Study. *Plasma Phys. Technol.* **2015**, *2*, 5–8.
3. Rümpler, C.; Narayanan, V. Arc Modeling Challenges. *Plasma Phys. Technol.* **2015**, *2*, 261–270.
4. McBride, J.W.; Balestrero, A.; Ghezzi, L.; Tribulato, G.; Cross, K.J. Optical fiber imaging for high speed plasma motion diagnostics: Applied to low voltage circuit breakers. *Rev. Sci. Instrum.* **2010**, *81*, 055109. [CrossRef] [PubMed]
5. Rigamonti, F. Diagnostic Methods for Electric Arc Plasma in Low Voltage Circuit Breakers: Modeling and Computational Aspects. Ph.D. Thesis, Politecnico di Milano, Milan, Italy, 2017.
6. Taccola, M. Diagnostic Methods for Electric Arc Plasma in Low Voltage Circuit Breakers: Measuring and Apparatus Design. Ph.D. Thesis, Politecnico di Milano, Milan, Italy, 2017.
7. Velleaud, G.; Mercier, M.; Laurent, A.; Gary, F. Use of an inverse method on the determination of the evolution of a self-blown electric arc in the air. *IEEE Trans. Plasma Sci.* **1991**, *19*, 510–514. [CrossRef]
8. Jackson, J.D. *Classical Electrodynamics*, 3rd ed.; John Wiley & Sons: Hoboken, NJ, USA, 1999.
9. Ripka, P.; Janosek, M. Advances in Magnetic Field Sensors. *IEEE Sens. J.* **2010**, *10*, 1108–1116. [CrossRef]
10. Lenz, J.E. A review of magnetic sensors. *Proc. IEEE* **1990**, *78*, 973–989. [CrossRef]
11. Honeywell. SS39ET/SS49ET/SS59ET Series. In *Linear Hall-Effect Sensor ICs, Technical Sheet*; Honeywell: Charlotte, NC, USA, 2012.
12. Groetsch, C.W. *Inverse Problems in the Mathematical Sciences*; Springer: Berlin/Heidelberg, Germany, 1993; Volume 52.
13. Nocedal, J.; Wright, S.J. *Numerical Optimization*; Springer: Berlin/Heidelberg, Germany, 1999.
14. Prando, S. The Magnetic Camera: Analysis and Improvement of Non-Invasive Short Circuit Diagnostics in Low Voltage Circuit Breakers. Master's Thesis, Politecnico di Milano, Milan, Italy, 2023.
15. Tarantola, A. *Inverse Problem Theory and Methods for Model Parameter Estimation*; SIAM: Philadelphia, PA, USA, 2005.
16. Idier, J. *Bayesian Approach to Inverse Problems*; John Wiley & Sons: Hoboken, NJ, USA, 2013.

MDPI AG
Grosspeteranlage 5
4052 Basel
Switzerland
Tel.: +41 61 683 77 34

Sensors Editorial Office
E-mail: sensors@mdpi.com
www.mdpi.com/journal/sensors

www.ingramcontent.com/pod-product-compliance
Lightning Source LLC
Chambersburg PA
CBHW061601210926
51850CB00009B/2173
* 9 7 8 3 7 2 5 8 5 5 6 3 6 *